Springer Undergraduate Mathematic

CW00420655

Springer

London
Berlin
Heidelberg
New York
Hong Kong
Milan
Paris
Tokyo

Advisory Board

Other books in this series

John M. Howie

Complex Analysis

With 83 Figures

 Springer

John M. Howie, CBE, MA, DPhil, DSc, Hon D.Univ., FRSE
School of Mathematics and Statistics, Mathematical Institute,
University of St Andrews, North Haugh, St Andrews, Fife, KY16 9SS, UK

Cover illustration elements reproduced by kind permission of:
Aptech Systems, Inc., Publishers of the GAUSS Mathematical and Statistical System, 23804 S.E. Kent-Kangley Road, Maple Valley, WA 98038, USA. Tel: (206) 432 - 7855 Fax (206) 432 - 7832 email: info@aptech.com URL: www.aptech.com
American Statistical Association: Chance Vol 8 No 1, 1995 article by KS and KW Heiner 'Tree Rings of the Northern Shawangunks' page 32 fig 2
Springer-Verlag: Mathematica in Education and Research Vol 4 Issue 3 1995 article by Roman E Maeder, Beatrice Amrhein and Oliver Gloor 'Illustrated Mathematics: Visualization of Mathematical Objects' page 9 fig 11, originally published as a CD ROM 'Illustrated Mathematics' by TELOS: ISBN 0-387-14222-3, German edition by Birkhauser: ISBN 3-7643-5100-4.
Mathematica in Education and Research Vol 4 Issue 3 1995 article by Richard J Gaylord and Kazume Nishidate 'Traffic Engineering with Cellular Automata' page 35 fig 2. Mathematica in Education and Research Vol 5 Issue 2 1996 article by Michael Trott 'The Implicitization of a Trefoil Knot' page 14.
Mathematica in Education and Research Vol 5 Issue 2 1996 article by Lee de Cola 'Coins, Trees, Bars and Bells: Simulation of the Binomial Process' page 19 fig 3. Mathematica in Education and Research Vol 5 Issue 2 1996 article by Richard Gaylord and Kazume Nishidate 'Contagious Spreading' page 33 fig 1. Mathematica in Education and Research Vol 5 Issue 2 1996 article by Joe Buhler and Stan Wagon 'Secrets of the Madelung Constant' page 50 fig 1.

British Library Cataloguing in Publication Data
Howie, John Mackintosh
    Complex analysis. – (Springer undergraduate mathematics
    Series)
    1.Mathematical analysis 2.Functions of complex variables
    I.Title
    515.9
    ISBN 1852337338

Library of Congress Cataloging-in-Publication Data
A catalog record for this book is available from the Library of Congress.

Springer Undergraduate Mathematics Series ISSN 1615-2085
ISBN 1-85233-733-8 Springer-Verlag London Berlin Heidelberg
Springer-Verlag is a part of Springer Science+Business Media
springeronline.com

Typesetting: Camera ready by author
12/3830-54321 Printed on acid-free paper SPIN 11012351

In memory of Katharine

Preface

Of all the central topics in the undergraduate mathematics syllabus, complex analysis is arguably the most attractive. The huge consequences emanating from the assumption of differentiability, and the sheer power of the methods deriving from Cauchy's Theorem never fail to impress, and undergraduates actively enjoy exploring the applications of the Residue Theorem.

Complex analysis is not an elementary topic, and one of the problems facing lecturers is that many of their students, particularly those with an "applied" orientation, approach the topic with little or no familiarity with the ϵ–δ arguments that are at the core of a serious course in analysis. It is, however, possible to appreciate the essence of complex analysis without delving too deeply into the fine detail of the proofs, and in the earlier part of the book I have starred some of the more technical proofs that may safely be omitted. Proofs are, however, given, since the development of more advanced analytical skills comes from imitating the techniques used in proving the major results.

The opening two chapters give a brief account of the preliminaries in real function theory and complex numbers that are necessary for the study of complex functions. I have included these chapters partly with self-study in mind, but they may also be helpful to those whose lecturers airily (and wrongly) assume that students remember everything learned in previous years.

In what is certainly designed as a first course in complex analysis I have deemed it appropriate to make only minimal reference to the topological issues that are at the core of the subject. This may be a disappointment to some professionals, but I am confident that it will be appreciated by the undergraduates for whom the book is intended.

The general plan of the book is fairly traditional, and perhaps the only slightly unusual feature is the brief final Chapter 12, which I hope will show that the subject is very much alive. In Section 12.2 I give a very brief and imprecise account of Julia sets and the Mandelbrot set, and in Section 12.1 I explain the Riemann Hypothesis, arguably the most remarkable and important unsolved problem in mathematics. If the eventual conqueror of the Riemann Hypothesis were to have learned the basics of complex analysis from this book, then I would rest content indeed!

All too often mathematics is presented in such a way as to suggest that it was engraved in pre-history on tablets of stone. The footnotes with the names and dates of the mathematicians who created complex analysis are intended to emphasise that mathematics was and is created by real people. Information on these people and their achievements can be found on the St Andrews website `www-history.mcs.st-and.ac.uk/history/`.

I am grateful to my colleague John O'Connor for his help in creating the diagrams. Warmest thanks are due also to Kenneth Falconer and Michael Wolfe, whose comments on the manuscript have, I hope, eliminated serious errors. The responsibility for any imperfections that remain is mine alone.

John M. Howie
University of St Andrews
January, 2003

Contents

1
What Do I Need to Know?

Introduction

Complex analysis is not an elementary subject, and the author of a book like this has to make some reasonable assumptions about what his readers know already. Ideally one would like to assume that the student has some basic knowledge of complex numbers and has experienced a fairly substantial first course in real analysis. But while the first of these requirements is realistic the second is not, for in many courses with an "applied" emphasis a course in complex analysis sits on top of a course on advanced (multi-variable) calculus, and many students approach the subject with little experience of ϵ–δ arguments, and with no clear idea of the concept of uniform convergence. This chapter sets out in summary the equipment necessary to make a start on this book, with references to suitable texts. It is written as a reminder: if there is anything you don't know at all, then at some point you will need to consult another book, either the suggested reference or another similar volume.

Given that the following summary might be a little indigestible, you may find it better to skip it at this stage, returning only when you come across anything unfamiliar. If you feel reasonably confident about complex numbers, then you might even prefer to skip Chapter 2 as well.

1.1 Set Theory

You should be familiar with the notations of set theory. See [9, Section 1.3].

If A is a set and a is a member, or **element**, of A we write $a \in A$, and if x is *not* an element of A we write $x \notin A$. If B is a subset of A we write $B \subseteq A$ (or sometimes $A \supseteq B$). If $B \subseteq A$ but $B \neq A$, then B is a **proper** subset of A. We write $B \subset A$, or $A \supset B$.

Among the subsets of A is the **empty set** $\emptyset$, containing no elements at all.

Sets can be described by listing, or by means of a defining property. Thus the set $\{3, 6, 9, 12\}$ (described by listing) can alternatively be described as $\{3x : x \in \{1, 2, 3, 4\}\}$ or as $\{x \in \{1, 2, \ldots, 12\} : 3 \text{ divides } x\}$.

The **union** $A \cup B$ of two sets is defined by:

$$x \in A \cup B \text{ if and only if } x \in A \text{ or } x \in B \text{ (or both)}.$$

The **intersection** $A \cap B$ is defined by

$$x \in A \cap B \text{ if and only if } x \in A \text{ and } x \in B.$$

The set $A \setminus B$ is defined by

$$A \setminus B = \{x \in A : x \notin B\}.$$

In the case where $B \subseteq A$ this is called the **complement** of B in A.

The **cartesian product** $A \times B$ of two sets A and B is defined by

$$A \times B = \{(a, b) : a \in A, b \in B\}.$$

1.2 Numbers

See [9, Section 1.1].

The following notations will be used:

$\mathbb{N} = \{1, 2, 3, \ldots\}$, the set of **natural numbers**;

$\mathbb{Z} = \{0, \pm 1, \pm 2, \ldots\}$, the set of **integers**;

$\mathbb{Q} = \{p/q : p, q \in \mathbb{Z}, q \neq 0\}$, the set of **rational numbers**;

$\mathbb{R}$, the set of **real numbers**.

It is not necessary to know any formal definition of $\mathbb{R}$, but certain properties are crucial. For each a in $\mathbb{R}$ the notation $|a|$, the **absolute value**, or **modulus**, of a, is defined by

$$|a| = \begin{cases} a & \text{if } a \geq 0 \\ -a & \text{if } a < 0. \end{cases}$$

If U is a subset of $\mathbb{R}$, then U is **bounded above** if there exists K in $\mathbb{R}$ such that $u \leq K$ for all u in U, and the number K is called an **upper bound** for U. Similarly, U is **bounded below** if there exists L in $\mathbb{R}$ such that $u \geq L$ for all u in U, and the number L is called a **lower bound** for U. The set U is **bounded** if it is bounded both above and below. Equivalently, U is bounded if there exists $M > 0$ such that $|u| \leq M$ for all u in U.

The **least upper bound** K for a set U is defined by the two properties

(i) K is an upper bound for U;

(ii) if K' is an upper bound for U, then $K' \geq K$.

The **greatest lower bound** is defined in an analogous way.

The **Least Upper Bound Axiom** for $\mathbb{R}$ states that *every non-empty subset of $\mathbb{R}$ that is bounded above has a least upper bound in $\mathbb{R}$*. Notice that the set $\mathbb{Q}$ does *not* have this property: the set $\{q \in \mathbb{Q} : q^2 < 2\}$ is bounded above, but has no least upper bound in $\mathbb{Q}$. It does of course have a least upper bound in $\mathbb{R}$, namely $\sqrt{2}$.

The least upper bound of a subset U is called the **supremum** of U, and is written $\sup U$. The greatest lower bound is called the **infimum** of U, and is written $\inf U$.

We shall occasionally use proofs by **induction**: if a proposition $\mathbb{P}(n)$ concerning natural numbers is true for $n = 1$, and if, for all $k \geq 1$ we have the implication $\mathbb{P}(k) \Longrightarrow \mathbb{P}(k+1)$, then $\mathbb{P}(n)$ is true for all n in $\mathbb{N}$. The other version of induction, sometimes called the Second Principle of Induction, is as follows: if $\mathbb{P}(1)$ is true and if, for all $m > 1$, the truth of $\mathbb{P}(k)$ for all $k < m$ implies the truth of $\mathbb{P}(m)$, then $\mathbb{P}(n)$ is true for all n.

One significant result that can be proved by induction (see [9, Theorem 1.7]) is

Theorem 1.1 (The Binomial Theorem)

For all a, b, and all integers $n \geq 1$,

$$(a + b)^n = \sum_{r=0}^{n} \binom{n}{r} a^{n-r} b^r .$$

Here
$$\binom{n}{r} = \frac{n!}{r!(n-r)!} = \frac{n(n-1)\ldots(n-r+1)}{r!} .$$

Note also the **Pascal Triangle Identity**

$$\binom{n}{r} + \binom{n}{r-1} = \binom{n+1}{r} . \tag{1.1}$$

EXERCISES

1.1. Show that the Least Upper Bound Axiom implies the **Greatest Lower Bound Axiom**: *every non-empty subset of* $\mathbb{R}$ *that is bounded below has a greatest lower bound in* $\mathbb{R}$.

1.2. Let the numbers $q_1, q_2, q_3, \ldots$ be defined by

$$q_1 = 2, \quad q_n = 3q_{n-1} - 1 \ (n \geq 2).$$

Prove by induction that

$$q_n = \frac{1}{2}(3^n + 1).$$

1.3. Let the numbers $f_1, f_2, f_3, \ldots$ be defined by

$$f_1 = f_2 = 1, \quad f_n = f_{n-1} + f_{n-2} \ (n \geq 3).$$

Prove by induction that

$$f_n = \frac{1}{\sqrt{5}}(\gamma^n - \delta^n),$$

where $\gamma = \frac{1}{2}(1 + \sqrt{5})$, $\delta = \frac{1}{2}(1 - \sqrt{5})$.

[This is the famous Fibonacci sequence. See [2].]

1.3 Sequences and Series

See [9, Chapter 2].

A sequence $(a_n)_{n \in \mathbb{N}}$, often written simply as (a_n), has a **limit** L if a_n can be made arbitrarily close to L for all sufficiently large n. More precisely, (a_n) has a limit L if, for all $\epsilon > 0$, there exists a natural number N such that $|a_n - L| < \epsilon$ for all $n > N$. We write $(a_n) \to L$, or $\lim_{n \to \infty} a_n = L$. Thus, for example, $((n+1)/n) \to 1$. A sequence with a limit is called **convergent**; otherwise it is **divergent**.

A sequence (a_n) is **monotonic increasing** if $a_{n+1} \geq a_n$ for all $n \geq 1$, and **monotonic decreasing** if $a_{n+1} \leq a_n$ for all $n \geq 1$. It is **bounded above** if there exists K such that $a_n \leq K$ for all $n \geq 1$. The following result is a key to many important results in real analysis:

Theorem 1.2

Every sequence (a_n) that is monotonic increasing and bounded above has a limit. The limit is $\sup\{a_n : n \geq 1\}$.

A sequence (a_n) is called a **Cauchy sequence**[1] if, for every $\epsilon > 0$, there exists a natural number N with the property that $|a_m - a_n| < \epsilon$ for all $m, n > N$. The **Completeness Property** of the set $\mathbb{R}$ is

Theorem 1.3

Every Cauchy sequence is convergent.

A **series** $\sum_{n=1}^{\infty} a_n$ determines a sequence (S_N) of **partial sums**, where $S_N = \sum_{n=1}^{N} a_n$. The series is said to **converge**, or to be **convergent**, if the sequence of partial sums is convergent, and $\lim_{N \to \infty} S_N$ is called the **sum to infinity**, or just the **sum**, of the series. Otherwise the series is **divergent**. The Completeness Property above translates for series into

Theorem 1.4 (The General Principle of Convergence)

If for every $\epsilon > 0$ there exists N such that

$$\left| \sum_{r=n+1}^{m} a_r \right| < \epsilon$$

for all $m > n > N$, then $\sum_{n=1}^{\infty} a_n$ is convergent.

For series $\sum_{n=1}^{\infty} a_n$ of positive terms there are two tests for convergence.

Theorem 1.5 (The Comparison Test)

Let $\sum_{n=1}^{\infty} a_n$ and $\sum_{n=1}^{\infty} x_n$ be series of positive terms.

(i) If $\sum_{n=1}^{\infty} a_n$ converges and if $x_n \leq a_n$ for all n, then $\sum_{n=1}^{\infty} x_n$ also converges.

(ii) If $\sum_{n=1}^{\infty} a_n$ diverges and if $x_n \geq a_n$ for all n, then $\sum_{n=1}^{\infty} x_n$ also diverges.

---

[1] Augustin-Louis Cauchy, 1789–1857.

Theorem 1.6 (The Ratio Test)

Let $\sum_{n=1}^{\infty} a_n$ be a series of positive terms.

(i) If $\lim_{n \to \infty}(a_{n+1}/a_n) = l < 1$, then $\sum_{n=1}^{\infty} a_n$ converges.

(ii) If $\lim_{n \to \infty}(a_{n+1}/a_n) = l > 1$, then $\sum_{n=1}^{\infty} a_n$ diverges.

In Part (i) of the Comparison Test it is sufficient to have $x_n \leq ka_n$ for some positive constant k, and it is sufficient also that the inequality should hold for all n exceeding some fixed number N. Similarly, in Part (ii) it is sufficient to have (for some fixed N) $x_n \geq ka_n$ for some positive constant k and for all $n > N$. In the Ratio Test it is important to note that no conclusion at all can be drawn if $\lim_{n \to \infty}(a_{n+1}/a_n) = 1$.

Theorem 1.7

The geometric series $\sum_{n=0}^{\infty} ar^n$ converges if and only if $|r| < 1$. Its sum is $a/(1-r)$.

Theorem 1.8

The series $\sum_{n=1}^{\infty}(1/n^k)$ is convergent if and only if $k > 1$.

A series $\sum_{n=1}^{\infty} a_n$ of positive and negative terms is called **absolutely convergent** if $\sum_{n=1}^{\infty} |a_n|$ is convergent. The convergence of $\sum_{n=1}^{\infty} |a_n|$ in fact implies the convergence of $\sum_{n=1}^{\infty} a_n$, and so every absolutely convergent series is convergent. The series is called **conditionally convergent** if $\sum_{n=1}^{\infty} a_n$ is convergent and $\sum_{n=1}^{\infty} |a_n|$ is not.

Theorem 1.9

For a **power series** $\sum_{n=0}^{\infty} a_n x^n$ there are three possibilities:

(a) the series converges for all x; or

(b) the series converges only for $x = 0$; or

(c) there exists a real number $R > 0$, called the **radius of convergence**, with the property that the series converges when $|x| < R$ and diverges when $|x| > R$.

We find it convenient to write $R = \infty$ in Case (a), and $R = 0$ in Case (b).

Two methods of finding the radius of convergence are worth recording here:

Theorem 1.10

Let $\sum_{n=0}^{\infty} a_n x^n$ be a power series. Then:

(i) the radius of convergence of the series is $\lim_{n\to\infty} |a_n/a_{n+1}|$, if this limit exists;

(ii) the radius of convergence of the series is $1/[\lim_{n\to\infty} |a_n|^{1/n}]$, if this limit exists.

We shall also encounter series of the form $\sum_{n=-\infty}^{\infty} b_n$. These cause no real difficulty, but it is important to realise that convergence of such a series requires the *separate* convergence of the two series $\sum_{n=0}^{\infty} b_n$ and $\sum_{n=1}^{\infty} b_{-n}$. It is *not* enough that $\lim_{N\to\infty} \sum_{n=-N}^{N} b_n$ should exist. Consider, for example, $\sum_{n=-\infty}^{\infty} n^3$, where $\sum_{n=-N}^{N} n^3 = 0$ for all N, but where it would be absurd to claim convergence.

1.4 Functions and Continuity

See [9, Chapter 3].

Let I be an interval, let $c \in I$, and let f be a real function whose domain $\operatorname{dom} f$ contains I, except possibly for the point c. We say that $\lim_{x\to c} f(x) = l$ if $f(x)$ can be made arbitrarily close to l by choosing x sufficiently close to c. More precisely, $\lim_{x\to c} f(x) = l$ if, for every $\epsilon > 0$, there exists $\delta > 0$ such that $|f(x) - l| < \epsilon$ for all x in $\operatorname{dom} f$ such that $0 < |x - c| < \delta$. If the domain of f contains c, we say that f is **continuous at** c if $\lim_{x\to c} f(x) = f(c)$. Also, f is **continuous on** I if it is continuous at every point in I.

The **exponential function** $\exp x$, often written e^x, is defined by the power series $\sum_{n=0}^{\infty} (x^n/n!)$. It has the properties

$$e^x > 0 \text{ for all } x, \quad e^{x+y} = e^x e^y, \quad e^{-x} = \frac{1}{e^x}.$$

The **logarithmic function** $\log x$, defined for $x > 0$, is the inverse function of e^x:

$$\log(e^x) = x \quad (x \in \mathbb{R}), \qquad e^{\log x} = x \quad (x > 0).$$

It has the properties

$$\log(xy) = \log x + \log y, \quad \log(1/x) = -\log x.$$

The following limits are important. (See [9, Section 6.3].)

$$\lim_{x \to \infty} x^\alpha e^{-kx} = 0 \quad (\alpha, k > 0);\tag{1.2}$$

$$\lim_{x \to \infty} x^{-k}(\log x)^\alpha = 0, \quad \lim_{x \to 0+} x^k(\log x)^\alpha = 0 \quad (\alpha, k > 0).\tag{1.3}$$

The **circular functions** cos and sin, defined by the series

$$\cos x = \sum_{n=0}^{\infty} (-1)^n \frac{x^{2n}}{(2n)!}, \quad \sin x = \sum_{n=0}^{\infty} (-1)^n \frac{x^{2n+1}}{(2n+1)!},\tag{1.4}$$

have the properties

$$\cos^2 x + \sin^2 x = 1,\tag{1.5}$$

$$\cos(-x) = \cos x \quad \sin(-x) = -\sin x,\tag{1.6}$$

$$\cos 0 = 1, \quad \sin 0 = 0, \quad \cos(\pi/2) = 0, \quad \sin(\pi/2) = 1,\tag{1.7}$$

$$\cos(x + y) = \cos x \cos y - \sin x \sin y,\tag{1.8}$$

$$\sin(x + y) = \sin x \cos y + \cos x \sin y.\tag{1.9}$$

All other identities concerning circular functions can be deduced from these, including the periodic properties

$$\cos(x + 2\pi) = \cos x, \quad \sin(x + 2\pi) = \sin x,$$

and the location of the zeros: $\cos x = 0$ if and only if $x = (2n + 1)\pi/2$ for some n in $\mathbb{Z}$; and $\sin x = 0$ if and only if $x = n\pi$ for some n in $\mathbb{Z}$.

The remaining circular functions are defined in terms of sin and cos as follows:

$$\tan x = \frac{\sin x}{\cos x}, \quad \sec x = \frac{1}{\cos x} \quad (x \neq (2n+1)\pi/2);$$

$$\cot x = \frac{\cos x}{\sin x}, \quad \operatorname{cosec} x = \frac{1}{\sin x} \quad (x \neq n\pi).$$

Remark 1.11

It is not obvious that the functions defined by the series (1.4) have any connection with the "adjacent over hypotenuse" and "opposite over hypotenuse" definitions one learns in secondary school. They are, however, the same. For an account, see [9, Chapter 8].

The inverse functions $\sin^{-1}$ and $\tan^{-1}$ need to be defined with some care. The domain of $\sin^{-1}$ is the interval $[-1, 1]$, and $\sin^{-1} x$ is the unique y in $[-\pi/2, \pi/2]$ such that $\sin y = x$. Then certainly $\sin(\sin^{-1} x) = x$ for all x in $[-1, 1]$, but we cannot say that $\sin^{-1}(\sin x) = x$ for all x in $\mathbb{R}$, for $\sin^{-1}(\sin x)$ must lie in the interval $[-\pi/2, \pi/2]$ whatever the value of x. Similarly, the domain of $\tan^{-1}$ is $\mathbb{R}$, and $\tan^{-1} x$ is defined as the unique y in the open interval $(-\pi/2, \pi/2)$ such that $\tan y = x$. Again, we have $\tan(\tan^{-1}(x)) = x$ for all x, but $\tan^{-1}(\tan x) = x$ only if $x \in (-\pi/2, \pi/2)$.

The **hyperbolic functions** are defined by

$$\cosh x = \frac{1}{2}(e^x + e^{-x}), \quad \sinh x = \frac{1}{2}(e^x - e^{-x}). \tag{1.10}$$

Equivalently,

$$\cosh x = \sum_{n=0}^{\infty} \frac{x^{2n}}{(2n)!}, \quad \sinh x = \sum_{n=0}^{\infty} \frac{x^{2n+1}}{(2n+1)!}. \tag{1.11}$$

By analogy with the circular functions, we define

$$\tanh x = \frac{\sinh x}{\cosh x}, \quad \operatorname{sech} x = \frac{1}{\cosh x} \quad (x \in \mathbb{R}),$$

$$\coth x = \frac{\cosh x}{\sinh x}, \quad \operatorname{cosech} x = \frac{1}{\sinh x} \quad (x \neq 0).$$

EXERCISES

1.4. Use the formulae $(1.5) - (1.9)$ to show that

$$\cos\frac{\pi}{4} = \sin\frac{\pi}{4} = \frac{1}{\sqrt{2}}.$$

1.5. a) Use the formulae $(1.5) - (1.9)$ to obtain the formula

$$\cos 3\theta = 4\cos^3\theta - 3\cos\theta,$$

and deduce that

$$\cos\frac{\pi}{6} = \frac{\sqrt{3}}{2}, \quad \sin\frac{\pi}{6} = \frac{1}{2}.$$

b) Hence show that

$$\cos\frac{\pi}{3} = \frac{1}{2}, \quad \sin\frac{\pi}{3} = \frac{\sqrt{3}}{2}.$$

1.6. Deduce from (1.8) and (1.9) that

$$\cos x + \cos y = 2\cos\frac{x+y}{2}\cos\frac{x-y}{2},$$
$$\cos x - \cos y = 2\sin\frac{x+y}{2}\sin\frac{y-x}{2}.$$

1.7. Define the sequence (a_n) by

$$a_1 = a_2 = 1 \quad a_n = 2a_{n-1} - 4a_{n-2} \ (n \geq 3).$$

Prove by induction that, for all $n \geq 1$,

$$a_n = 2^{n-1}\cos\frac{(n-1)\pi}{3}.$$

1.5 Differentiation

See [9, Chapter 4].

A function f is **differentiable** at a point a in its domain if the limit

$$\lim_{x \to a} \frac{f(x) - f(a)}{x - a}$$

exists. The value of the limit is called the **derivative** of f at a, and is denoted by $f'(a)$. A function is **differentiable** in an interval (a, b) if it is differentiable at every point in (a, b).

The function $f'(x)$ is alternatively denoted by

$$\frac{d}{dx}[f(x)], \quad \text{or} \quad (D_x f)(x) \quad \text{or} \quad \frac{dy}{dx},$$

where $y = f(x)$.

Theorem 1.12 (The Mean Value Theorem)

If f is continuous in $[a, b]$ and differentiable in (a, b), and if $x \in (a, b)$, then there exists u in (a, b) such that

$$f(x) = f(a) + (x - a)f'(u).$$

Moreover, if f' exists and is continuous in $[a, b]$, then

$$f(x) = f(a) + (x - a)\big(f'(a) + \epsilon(x)\big),$$

where $\epsilon(x) \to 0$ as $x \to a$.

Corollary 1.13

Let f be continuous in $[a, b]$ and differentiable in (a, b), and suppose that $f'(x) = 0$ for all x in (a, b). Then f is a constant function.

The following table of functions and derivatives may be a useful reminder:

$f(x)$	$f'(x)$
x^n	nx^{n-1}
e^x	e^x
$\log x \quad (x > 0)$	$1/x$
$\sin x$	$\cos x$
$\cos x$	$-\sin x$
$\tan x \quad (x \neq (n + \frac{1}{2})\pi)$	$1/\cos^2 x$
$\sin^{-1} x \quad (|x| < 1)$	$1/\sqrt{1 - x^2}$
$\tan^{-1} x$	$1/(1 + x^2)$

Recall also the crucial techniques of differential calculus. Here u and v are differentiable in some interval containing x.

The Linearity Rule. If $f(x) = ku(x) + lv(x)$, where k, l are constants, then

$$f'(x) = ku'(x) + lv'(x).$$

The Product Rule. If $f(x) = u(x)v(x)$, then

$$f'(x) = u'(x)v(x) + u(x)v'(x).$$

The Quotient Rule. If $f(x) = u(x)/v(x)$ (where $v(x) \neq 0$) then

$$f'(x) = \frac{v(x)u'(x) - u(x)v'(x)}{[v(x)]^2}.$$

The Chain Rule. If $f(x) = u\big(v(x)\big)$, then

$$f'(x) = u'\big(v(x)\big).v'(x).$$

We shall have cause to deal with **higher derivatives** also. A function f may have a derivative f' that is differentiable, and in this case we denote the derivative of f' by f''. The process can continue: we obtain derivatives $f''', f^{(4)}, \ldots, f^{(n)}, \ldots$. (Obviously the transition from dashes to bracketed superscripts is a bit arbitrary: if we write "$f^{(n)}$ $(n \geq 0)$", then by $f^{(0)}$, $f^{(1)}$, $f^{(2)}$ and $f^{(3)}$ we mean (respectively) f, f', f'' and f'''.) The linearity rule applies without change to higher derivatives, and the product rule is replaced by **Leibniz's**[2] **Theorem**:

Theorem 1.14

Let f, g be functions that are n times differentiable. Then

$$(f \cdot g)^{(n)} = \sum_{r=0}^{n} \binom{n}{r} f^{(n-r)} \cdot g^{(r)} \, .$$

1.6 Integration

See [9, Chapter 5].

It is not necessary to have studied any formal integration theory, but you should know the following results.

Theorem 1.15 (The Fundamental Theorem of Calculus)

Let f be continuous in $[a, b]$, and let

$$F(x) = \int_a^x f(u) \, du \quad \big(x \in [a, b]\big) \, .$$

Then F is differentiable in $[a, b]$, and $F'(x) = f(x)$.

Theorem 1.16 (The Antiderivative Theorem)

Let f be continuous in $[a, b]$. Then there exists a function Φ such that $\Phi'(x) = f(x)$ for all x in $[a, b]$, and

$$\int_a^b f(x) \, dx = \Phi(b) - \Phi(a) \, .$$

---

[2] Gottfried Wilhelm Leibniz, 1646–1716.

Of course, the *existence* of the function Φ does not mean than we can express it in terms of functions we know. The table of derivatives in the last section gives rise to a corresponding table of antiderivatives:

$f(x)$	$\Phi(x)$
$x^n \quad (n \neq -1)$	$x^{n+1}/(n+1)$
e^x	e^x
$1/x$	$\log|x|$
$\sin x$	$-\cos x$
$\cos x$	$\sin x$
$1/\sqrt{1-x^2} \quad (|x| < 1)$	$\sin^{-1} x$
$1/(1+x^2)$	$\tan^{-1} x$

We usually denote an antiderivative of f by $\int f(x)\,dx$. It is defined only to within an arbitrary constant – by which we mean that, if $\Phi(x)$ is an antiderivative, then so is $\Phi(x) + C$, where C is an arbitrary constant.

The finding of antiderivatives is intrinsically harder than the finding of derivatives. Corresponding to the Linearity Rule, the Product Rule and the Chain Rule for differentiation we have

The Linearity Rule. If $f(x) = ku(x) + lv(x)$, where k, l are constants, then

$$\int f(x)\,dx = k \int u(x)\,dx + l \int v(x)\,dx.$$

Integration by Parts.

$$\int u(x)v(x)\,dx = u(x) \int v(x)\,dx - \int u'(x)\left[\int v(x)\,dx\right] dx.$$

Integration by Substitution. Let f be continuous in $[a, b]$, and let g be a function whose derivative g' is either positive throughout, or negative throughout, the interval $[a, b]$. Then

$$\int_a^b f(x)\,dx = \int_{g^{-1}(a)}^{g^{-1}(b)} f\big(g(u)\big)g'(u)\,du.$$

This last rule looks more frightening than it is, and here the dy/dx notation for derivatives is useful. To evaluate

$$I = \int_0^1 \frac{x^5}{\sqrt{1+x^6}}\,dx$$

we argue as follows. Let $u = 1 + x^6$. Then $du = 6x^5 \, dx$. Also, $x = 0$ gives $u = 1$, and $x = 1$ gives $u = 2$. So

$$I = \int_1^2 \frac{1}{6} u^{-1/2} \, du = \left[\frac{1}{3} u^{1/2} \right]_1^2 = \frac{1}{3}(\sqrt{2} - 1).$$

(Here $g(u) = (u - 1)^{1/6}$, $g^{-1}(x) = 1 + x^6$.)

1.7 Infinite Integrals

See [9, Sections 5.6 and 5.7].

If $\lim_{K \to \infty} \int_a^K f(x) \, dx$ exists, we write the limit as $\int_a^\infty f(x) \, dx$, and say that the integral is **convergent**. Similarly, we write $\lim_{L \to \infty} \int_{-L}^a f(x) \, dx$, if it exists, as $\int_{-\infty}^a f(x) \, dx$. If both limits exist, we write

$$\int_{-\infty}^\infty f(x) \, dx = \int_{-\infty}^a f(x) \, dx + \int_a^\infty f(x) \, dx.$$

(The value is, of course, independent of a.)

It is important to note that, by analogy with sums from $-\infty$ to ∞, we require the *separate* existence of $\int_{-\infty}^a f(x) \, dx$ and $\int_a^\infty f(x) \, dx$, and that this is a stronger requirement than the existence of $\lim_{K \to \infty} \int_{-K}^K f(x) \, dx$. The latter limit exists, for example, for any odd function $(f(-x) = -f(x))$ whatever.

The limit $\lim_{K \to \infty} \int_{-K}^K f(x) \, dx$ is often called the **Cauchy principal value** of $\int_{-\infty}^\infty f(x) \, dx$, and is written $(PV) \int_{-\infty}^\infty f(x) \, dx$. If $\int_{-\infty}^\infty f(x) \, dx$ is convergent, then

$$(PV) \int_{-\infty}^\infty f(x) \, dx = \int_{-\infty}^\infty f(x) \, dx.$$

The theory of infinite integrals closely parallels the theory of infinite series, and, as with infinite series, we say that $\int_a^\infty f(x) \, dx$ is **absolutely convergent** if $\int_a^\infty |f(x)| \, dx$ is convergent. As with series, absolute convergence implies convergence.

By analogy with Theorem 1.8, we have

Theorem 1.17

The integrals

$$\int_1^\infty \frac{dx}{x^n} \quad \text{and} \quad \int_{-\infty}^{-1} \frac{dx}{x^n}$$

are convergent if and only if $n > 1$.

Given two functions f and g, we write $f \asymp g$ (as $x \to \infty$) (read "f has the same order of magnitude as g") if

$$\lim_{x \to \infty} \frac{f(x)}{g(x)} = K,$$

where $K > 0$. The stronger statement $f \sim g$ (read "f is asymptotically equal to g") means that

$$\lim_{x \to \infty} \frac{f(x)}{g(x)} = 1.$$

This gives us the most convenient form of the Comparison Test:

Theorem 1.18

Let f, g be positive bounded functions on $[a, \infty)$ such that $f \asymp g$. Then $\int_a^\infty f(x)\, dx$ is convergent if and only if $\int_a^\infty g(x)\, dx$ is convergent.

Example 1.19

Show that

$$\int_{-\infty}^{\infty} \frac{dx}{(1 + x^6)^{1/3}}$$

is convergent.

Solution

Compare the integrand with $1/(1 + x^2)$. Since $(1 + x^2)/(1 + x^6)^{1/3} \to 1$ as $x \to \pm\infty$, both the integral from 0 to ∞ and the integral from $-\infty$ to 0 exist, by Theorems 1.18 and 1.17. □

We often encounter integrals in which the integrand has a singularity somewhere in the range. If $f(x) \to \infty$ as $x \to a+$, but f is bounded in any $[a + \epsilon, b]$, we say that the **improper** integral $\int_a^b f(x)\, dx$ exists if $\int_{a+\epsilon}^b f(x)\, dx$ has a finite limit as $\epsilon \to 0+$. A similar definition applies if $f(x) \to \infty$ as $x \to b-$. If the singularity is at c, where $a < c < b$, then the integral exists only if *both* the limits

$$\lim_{\eta \to 0+} \int_a^{c-\eta} f(x)\, dx \quad \text{and} \quad \lim_{\epsilon \to 0+} \int_{c+\epsilon}^b f(x)\, dx$$

exist, and $\int_a^b f(x)\, dx$ is the sum of these limits. This is a stronger requirement than the existence of

$$\lim_{\epsilon \to 0+} \left[\int_a^{c-\epsilon} f(x)\, dx + \int_{c+\epsilon}^b f(x)\, dx \right].$$

If this latter limit exists we have a **Cauchy principal value**, and we denote the limit by $(PV) \int_a^b f(x)\, dx$. For example, $(PV) \int_{-1}^1 (1/x)\, dx = 0$, but $\int_{-1}^1 (1/x)\, dx$ does not exist.

In fact we have a pair of theorems that can often enable us to establish the convergence of improper integrals. First, it is easy to prove the following result.

Theorem 1.20

The integral

$$\int_a^b \frac{dx}{(b-x)^n}$$

converges if and only if $n < 1$. The integral

$$\int_a^b \frac{dx}{(x-a)^n}$$

converges if and only if $n < 1$.

It is straightforward to modify the definitions of $\asymp$ and $\sim$ to deal with limits as $x \to c$, where c is a real number. Thus $f \asymp g$ as $x \to c$ if there exists K in $\mathbb{R}$ such that $\lim_{x \to c}[f(x)/g(x)] = K$, and $f \sim g$ as $x \to c$ if $\lim_{x \to c}[f(x)/g(x)] = 1$.

Theorem 1.21

Suppose that f is bounded in $[a, b]$ except that $f(x) \to \infty$ as $x \to b-$. If $f(x) \asymp 1/(b-x)^n$ as $x \to b$, then $\int_a^b f(x)\, dx$ converges if and only if $n < 1$.

A similar conclusion exists if the singularity of the integrand is as the lower end of the range of integration.

Example 1.22

Show that

$$\int_0^1 \frac{dx}{\sqrt{1-x^2}}$$

converges.

Solution

Here $f(x) \to \infty$ as $x \to 1$. Now

$$\lim_{x \to 1-} \frac{1-x}{1-x^2} = \lim_{x \to 1-} \frac{1}{1+x} = \frac{1}{2},$$

and so

$$\lim_{x \to 1-} \frac{1/\sqrt{1-x^2}}{1/(1-x)^{1/2}} = \frac{1}{\sqrt{2}}.$$

Thus $f(x) \asymp 1/(1-x)^{1/2}$ as $x \to 1-$, and so the integral converges. □

1.8 Calculus of Two Variables

Let $f : (x,y) \mapsto f(x,y)$ be a function from $\mathbb{R} \times \mathbb{R}$ into $\mathbb{R}$. Then we say that $\lim_{(x,y) \to (a,b)} f(x,y) = L$ if, for all $\epsilon > 0$, there exists $\delta > 0$ such that $|f(x,y) - L| < \epsilon$ whenever $0 < \sqrt{(x-a)^2 + (y-b)^2} < \delta$. The function f is **continuous** at (a,b) if $\lim_{(x,y) \to (a,b)} f(x,y) = f(a,b)$. These definitions are in essence the same as for functions of a single variable: the distance $|x-a|$ between two points x and a in $\mathbb{R}$ is replaced by the distance $\sqrt{(x-a)^2 + (y-b)^2}$ between two points (x,y) and (a,b) in $\mathbb{R}^2$. [Note that when we write $\sqrt{}$ we always mean the *positive* square root.]

The limit

$$\lim_{h \to 0} \frac{f(a+h,b) - f(a,b)}{h},$$

if it exists, is called the **partial derivative of f with respect to x at (a,b)**, and is denoted by

$$\frac{\partial f}{\partial x}, \text{ or } \frac{\partial f}{\partial x}(a,b), \text{ or } f_x(a,b).$$

Similarly,

$$\lim_{k \to 0} \frac{f(a,b+k) - f(a,b)}{k}$$

is the **partial derivative of f with respect to y at (a,b)**, and is denoted by

$$\frac{\partial f}{\partial y}, \text{ or } \frac{\partial f}{\partial y}(a,b), \text{ or } f_y(a,b).$$

By analogy with the familiar notation in one-variable calculus, we write

$$\frac{\partial}{\partial x}\left(\frac{\partial f}{\partial x}\right) \text{ as } \frac{\partial^2 f}{\partial x^2}, \quad \frac{\partial}{\partial y}\left(\frac{\partial f}{\partial y}\right) \text{ as } \frac{\partial^2 f}{\partial y^2}, \quad \text{and } \frac{\partial}{\partial x}\left(\frac{\partial f}{\partial y}\right) \text{ as } \frac{\partial^2 f}{\partial x \partial y}.$$

Alternative notations are f_{xx}, f_{yy} and f_{xy}.

Suppose that $f : \mathbb{R} \times \mathbb{R} \to \mathbb{R}$ is a function whose partial derivatives are continuous. Then

$$f(a+h, b+k) = f(a,b) + h\big(f_x(a,b) + \epsilon_1\big) + k\big(f_y(a,b) + \epsilon_2\big), \qquad (1.12)$$

where $\epsilon_1, \epsilon_2 \to 0$ as $h, k \to 0$.

Suppose now that $z = f(u, v)$, where u and v are functions of x and y. The **chain rule** for functions of two variables is

$$\frac{\partial z}{\partial x} = \frac{\partial z}{\partial u}\frac{\partial u}{\partial x} + \frac{\partial z}{\partial v}\frac{\partial v}{\partial x}, \quad \frac{\partial z}{\partial y} = \frac{\partial z}{\partial u}\frac{\partial u}{\partial y} + \frac{\partial z}{\partial v}\frac{\partial v}{\partial y}.$$

2

Complex Numbers

2.1 Are Complex Numbers Necessary?

Much of mathematics is concerned with various kinds of equations, of which equations with numerical solutions are the most elementary. The most funda-mental set of numbers is the set $\mathbb{N} = \{1, 2, 3, \ldots\}$ of **natural numbers**. If a and b are natural numbers, then the equation $x + a = b$ has a solution *within the set of natural numbers* if and only if $a < b$. If $a \geq b$ we must *extend* the number system to the larger set $\mathbb{Z} = \{\ldots, -2, -1, 0, 1, 2, 3 \ldots\}$ of **integers**. Here we get a bonus, for the equation $x + a = b$ has a solution $x = b - a$ in $\mathbb{Z}$ *for all a and b in $\mathbb{Z}$.*

If $a, b \in \mathbb{Z}$ and $a \neq 0$, then the equation $ax + b = 0$ has a solution in $\mathbb{Z}$ if and only if a *divides* b. Otherwise we must once again extend the number system to the larger set $\mathbb{Q}$ of **rational numbers**. Once again we get a bonus, for the equation $ax + b = 0$ has a solution $x = -b/a$ in $\mathbb{Q}$ for all $a \neq 0$ in $\mathbb{Q}$ and all b in $\mathbb{Q}$.

When we come to consider a **quadratic** equation $ax^2 + bx + c = 0$ (where $a, b, c \in \mathbb{Q}$ and $a \neq 0$) we encounter our first real difficulty. We may safely assume that a, b and c are integers: if not, we simply multiply the equation by a suitable positive integer. The standard solution to the equation is given by the familiar formula

$$x = \frac{-b \pm \sqrt{b^2 - 4ac}}{2a}.$$

Let us denote $b^2 - 4ac$, the **discriminant** of the equation, by Δ. If Δ is the square of an integer (what is often called a **perfect square**) then the equation has rational solutions, and if Δ is positive then the two solutions are in the extended set $\mathbb{R}$ of real numbers. But if $\Delta < 0$ then there is no solution even within $\mathbb{R}$.

We have already carried out three extensions (to $\mathbb{Z}$, to $\mathbb{Q}$, to $\mathbb{R}$) from our starting point in natural numbers, and there is no reason to stop here. We can modify the standard formula to obtain

$$x = \frac{-b \pm \sqrt{(-1)(4ac - b^2)}}{2a} \, ,$$

where $4ac - b^2 > 0$. If we postulate the existence of $\sqrt{-1}$, then we get a "solution"

$$x = \frac{-b \pm \sqrt{-1}\sqrt{4ac - b^2}}{2a} \, .$$

Of course we know that there is no real number $\sqrt{-1}$, but the idea seems in a way to work. If we look at a specific example,

$$x^2 + 4x + 13 = 0 \, ,$$

and decide to write i for $\sqrt{-1}$, the formula gives us two solutions $x = -2 + 3i$ and $x = -2 - 3i$. If we use normal algebraic rules, replacing i^2 by -1 whenever it appears, we find that

$$(-2 + 3i)^2 + 4(-2 + 3i) + 13 = (-2)^2 + 2(-2)(3i) + (3i)^2 - 8 + 12i + 13$$
$$= 4 - 12i - 9 - 8 + 12i + 13 \text{ (since } i^2 = -1)$$
$$= 0 \, ,$$

and the validity of the other root can be verified in the same way. We can certainly agree that if there is a number system containing "numbers" $a + bi$, where $a, b \in \mathbb{R}$, then they will add and multiply according to the rules

$$(a_1 + b_1 i) + (a_2 + b_2 i) = (a_1 + a_2) + (b_1 + b_2)i \qquad (2.1)$$
$$(a_1 + b_1 i)(a_2 + b_2 i) = (a_1 a_2 - b_1 b_2) + (a_1 b_2 + b_1 a_2)i \, . \qquad (2.2)$$

We shall see shortly that there is a way, closely analogous to our picture of real numbers as points on a line, of visualising these new **complex** numbers.

Can we find equations that require us to extend our new complex number system (which we denote by $\mathbb{C}$) still further? No, in fact we cannot: the important **Fundamental Theorem of Algebra**, which we shall prove in Chapter 7, states that, for all $n \geq 1$, every polynomial equation

$$a_n x^n + a_{n-1} x^{n-1} + \cdots + a_1 x + a_0 = 0 \, ,$$

with coefficients $a_0, a_1, \ldots, a_n$ in $\mathbb{C}$ and $a_n \neq 0$, has all its roots within $\mathbb{C}$. This is one of many reasons why the number system $\mathbb{C}$ is of the highest importance in the development and application of mathematical ideas.

EXERCISES

2.1. One way of proving that the set $\mathbb{C}$ "exists" is to define it as the set of all 2×2 matrices

$$M(a, b) = \begin{pmatrix} a & b \\ -b & a \end{pmatrix},$$

where $a, b \in \mathbb{R}$.

a) Determine the sum and product of $M(a, b)$ and $M(c, d)$.

b) Show that

$$M(a, 0) + M(b, 0) = M(a + b, 0), \quad M(a, 0)M(b, 0) = M(ab, 0).$$

Thus $\mathbb{C}$ contains the real numbers "in disguise" as 2×2 diagonal matrices. Identify $M(a, 0)$ with the real number a.

c) With this identification, show that $M(0, 1)^2 = -1$. Denote $M(0, 1)$ by i.

d) Show that $M(a, b) = a + bi$.

2.2. Determine the roots of the equation $x^2 - 2x + 5 = 0$.

2.2 Basic Properties of Complex Numbers

We can visualise a complex number $z = x + yi$ as a point (x, y) on the plane. Real numbers x appear as points $(x, 0)$ on the x-axis, and numbers yi as points $(0, y)$ on the y-axis. Numbers yi are often called **pure imaginary**, and for this reason the y-axis is called the **imaginary axis**. The x-axis, for the same reason, is referred to as the **real axis**. It is important to realise that these terms are used for historical reasons only: within the set $\mathbb{C}$ the number $3i$ is no more "imaginary" than the number 3.

If $z = x + iy$, where x and y are real, we refer to x as the **real part** of z and write $x = \operatorname{Re} z$. Similarly, we refer to y as the **imaginary part** of z, and write $y = \operatorname{Im} z$. Notice that *the imaginary part of z is a real number*.

The number $\bar{z} = x - iy$ is called the **conjugate** of z. It is easy to verify that, for all complex numbers z and w,

$$\bar{\bar{z}} = z, \quad \overline{z + w} = \bar{z} + \bar{w}, \quad \overline{zw} = \bar{z}\bar{w}, \tag{2.3}$$

and

$$z + \bar{z} = 2\operatorname{Re} z, \quad z - \bar{z} = 2i\operatorname{Im} z \tag{2.4}$$

Note also that $\bar{z} = z$ if and only if z is real, and $\bar{z} = -z$ if and only if z is pure imaginary.

The following picture of a complex number $c = a + ib$ is very useful.

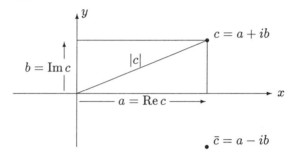

The product $c\bar{c}$ is the non-negative real number $a^2 + b^2$. Its square root $\sqrt{c\bar{c}} = \sqrt{a^2 + b^2}$, the distance of the point (a, b) from the origin, is denoted by $|c|$ and is called the **modulus** of c. If c is real, then the modulus is simply the **absolute value** of c. Some of the following results are familiar in the context of real numbers:

Theorem 2.1

Let z and w be complex numbers. Then:

(i) $|\operatorname{Re} z| \leq |z|$, $|\operatorname{Im} z| \leq |z|$, $|\bar{z}| = |z|$;

(ii) $|zw| = |z||w|$;

(iii) $|z + w| \leq |z| + |w|$;

(iv) $|z - w| \geq ||z| - |w||$.

Proof

(i) is immediate.

(ii) By (2.3),
$$|zw|^2 = (zw)(\overline{zw}) = (z\bar{z})(w\bar{w}) = \big(|z||w|\big)^2$$
and Part (ii) follows immediately.

For Part (iii), observe that

$$|z + w|^2 = (z + w)(\bar{z} + \bar{w}) = z\bar{z} + z\bar{w} + w\bar{z} + w\bar{w}. \qquad (2.5)$$

Now,

$$z\bar{w} + w\bar{z} = z\bar{w} + \overline{z\bar{w}} = 2\operatorname{Re}(z\bar{w}) \leq 2|z\bar{w}| = 2|z||\bar{w}| = 2|z||w|,$$

and so, from (2.5),

$$|z + w|^2 \leq |z|^2 + 2|z||w| + |w|^2 = (|z| + |w|)^2 \,.$$

The result now follows by taking square roots.

For Part (iv), we observe first that

$$|z| = \big|(z - w) + w\big| \leq |z - w| + |w|$$

and deduce that

$$|z - w| \geq |z| - |w| \,. \tag{2.6}$$

Similarly, from

$$|w| = \big|z - (z - w)\big| \leq |z| + |z - w|$$

we deduce that

$$|z - w| \geq |w| - |z| \,. \tag{2.7}$$

Hence, since for a real number x we have that $|x| = \max\{x, -x\}$, we deduce from (2.6) and (2.7) that

$$|z - w| \geq \big||z| - |w|\big| \,.$$

$\square$

The correspondence between complex numbers $c = a + bi$ and points (a, b) in the plane is so close that we shall routinely refer to "the point c", and we shall refer to the plane as the **complex plane**, or as the **Argand**[1] **diagram**. The point c lies on the circle $x^2 + y^2 = r^2$, where $r = |c| = \sqrt{a^2 + b^2}$.

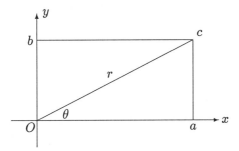

If $c \neq 0$ there is a unique θ in the interval $(-\pi, \pi]$ such that

$$\cos\theta = \frac{a}{\sqrt{a^2 + b^2}} \,, \quad \sin\theta = \frac{b}{\sqrt{a^2 + b^2}} \,,$$

and we can write

$$c = r(\cos\theta + i\sin\theta) \,. \tag{2.8}$$

_____

[1] Jean-Robert Argand, 1768–1822.

This amounts to describing the point (a, b) by means of **polar coordinates**, and $r(\cos\theta + i\sin\theta)$ is called the **polar form** of the complex number. By some standard trigonometry we see that

$$
\begin{aligned}
&(\cos\theta + i\sin\theta)(\cos\phi + i\sin\phi) \\
&= (\cos\theta\cos\phi - \sin\theta\sin\phi) + i(\sin\theta\cos\phi + \cos\theta\sin\phi) \\
&= \cos(\theta + \phi) + i\sin(\theta + \phi)\,.
\end{aligned}
\tag{2.9}
$$

Looking ahead to a notation that we shall justify properly in Chapter 3, we note that, if we extend the series definition of the exponential function to complex numbers, we have, for any real θ,

$$
\begin{aligned}
e^{i\theta} &= 1 + i\theta + \frac{1}{2!}(i\theta)^2 + \frac{1}{3!}(i\theta)^3 + \frac{1}{4!}(i\theta)^4 + \frac{1}{5!}(i\theta)^5 + \cdots \\
&= \left(1 - \frac{1}{2!}\theta^2 + \frac{1}{4!}\theta^4 - \cdots\right) + i\left(\theta - \frac{1}{3!}\theta^3 + \frac{1}{5!}\theta^5 - \cdots\right) \\
&= \cos\theta + i\sin\theta\,.
\end{aligned}
$$

We may therefore write (2.9) in the easily remembered form

$$
e^{i\theta}e^{i\phi} = e^{i(\theta+\phi)}\,.
$$

From well known properties of sin and cos we deduce that

$$
e^{i(-\theta)} = \cos(-\theta) + i\sin(-\theta) = \cos\theta - i\sin\theta\,,
$$

and **Euler's**[2] **formulae** follow easily:

$$
\cos\theta = \frac{1}{2}(e^{i\theta} + e^{-i\theta})\,, \quad \sin\theta = \frac{1}{2i}(e^{i\theta} - e^{-i\theta})\,.
\tag{2.10}
$$

With the exponential notation, the polar form for the non-zero complex number $c = a + bi$ is written as $re^{i\theta}$, where $a = r\cos\theta$, $b = r\sin\theta$. The positive number r is the **modulus** $|c|$ of c, and θ is the **argument**, written $\arg c$, of c. The polar form of $\bar{c}$ is $re^{-i\theta}$.

The periodicity of sin and cos implies that $e^{i\theta} = e^{i(\theta+2n\pi)}$ for every integer n, and so, more precisely, we specify $\arg c$ by the property that $\arg c = \theta$, where $c = re^{i\theta}$ and $-\pi < \theta \le \pi$. We call $\arg c$ the **principal argument** if there is any doubt.

Multiplication for complex numbers is easy if they are in polar form:

$$
(r_1 e^{i\theta_1})(r_2 e^{i\theta_2}) = (r_1 r_2)e^{i(\theta_1 + \theta_2)}\,.
$$

We already know that $|c_1 c_2| = |c_1|\,|c_2|$, and we now deduce that

$$
\arg(c_1 c_2) \equiv \arg c_1 + \arg c_2 \pmod{2\pi}\,.
$$

---

[2] Leonhard Euler, 1707–1783.

By this we mean that the difference between $\arg(c_1 c_2)$ and $\arg c_1 + \arg c_2$ is an integral multiple of 2π.

The results extend:

$$|c_1 c_2 \ldots c_n| = |c_1| \, |c_2| \ldots |c_n|,$$

$$\arg c_1 c_2 \ldots c_n \equiv \arg c_1 + \arg c_2 + \cdots + \arg c_n \quad (\mathrm{mod} \ 2\pi),$$

and, putting $c_1 = c_2 = \cdots = c_n$, we also deduce that, for all positive integers n,

$$|c^n| = |c|^n, \quad \arg c^n \equiv n \arg c \quad (\mathrm{mod} \ 2\pi).$$

Example 2.2

Determine the modulus and argument of c^5, where $c = 1 + i\sqrt{3}$.

Solution

An easy calculation gives $|c| = 2$, $\arg c = \theta$, where $\cos\theta = 1/2$, $\sin\theta = \sqrt{3}/2$; hence $\arg\theta = \pi/3$. It follows that $|c^5| = 2^5 = 32$, while $\arg(c^5) \equiv 5\pi/3 \equiv -\pi/3$. (Here we needed to make an adjustment in order to arrive at the principal argument.) The **standard form** of c^5, by which we mean the form $a + ib$, where a and b are real, is $32\big(\cos(-\pi/3) + i\sin(-\pi/3)\big) = 16(1 - i\sqrt{3})$. $\square$

Remark 2.3

For a complex number $c = a + bi = re^{i\theta}$ it is true that $\tan\theta = b/a$, but it is *not* always true that $\theta = \tan^{-1}(b/a)$. For example, if $c = -1 - i$, then $\theta = -3\pi/4 \neq \tan^{-1} 1$. It is much safer – indeed essential – to find θ by using $\cos\theta = a/r$, $\sin\theta = b/r$.

Finding the reciprocal of a non-zero complex number c is again easy if the number is in polar form: the reciprocal of $re^{i\theta}$ is $(1/r)e^{-i\theta}$. In the standard form $c = a + bi$ the reciprocal is less obvious:

$$\frac{1}{a + bi} = \frac{a - bi}{a^2 + b^2}.$$

The technique of multiplying the denominator of a fraction by its conjugate is worth noting:

Example 2.4

Express

$$\frac{3+7i}{2+5i}$$

in standard form.

Solution

$$\frac{3+7i}{2+5i} = \frac{(3+7i)(2-5i)}{(2+5i)(2-5i)} = \frac{1}{29}(41-i).$$

$\square$

Again, the fact that every complex number has a square root is easily seen from the polar form: $\sqrt{r}e^{i(\theta/2)}$ is a square root of $re^{i\theta}$. From this we may deduce that every quadratic equation

$$az^2 + bz + c = 0,$$

where $a, b, c \in \mathbb{C}$ and $a \neq 0$ has a solution in $\mathbb{C}$. The procedure, by "completing the square", and the resulting formula

$$z = \frac{-b \pm \sqrt{b^2 - 4ac}}{2a}$$

are just the same as for real quadratic equations.

Example 2.5

Find the roots of the equation

$$z^2 + 2iz + (2 - 4i) = 0.$$

Solution

By the standard formula, the solution of the equation is

$$\frac{1}{2}\left(-2i \pm \sqrt{(-2i)^2 - 4(2 - 4i)}\right) = \frac{1}{2}\left(-2i \pm \sqrt{-12 + 16i}\right) = -i \pm \sqrt{-3 + 4i}.$$

Observe now that $(1 + 2i)^2 = -3 + 4i$, and so the solution is

$$z = -i \pm (1 + 2i) = 1 + i \quad \text{or} \quad -1 - 3i.$$

$\square$

The addition of complex numbers has a strong geometrical connection, being in effect vector addition:

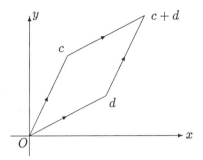

The geometrical aspect of complex multiplication becomes apparent if we use the polar form: if we multiply c by $re^{i\theta}$ we multiply $|c|$ by a factor of r, and add θ to $\arg c$. Of special interest is the case where $r = 1$, when multiplication by $e^{i\theta}$ corresponds simply to a rotation by θ. In particular:

– multiplication by $-1 = e^{i\pi}$ rotates by π;

– multiplication by $i = e^{i\pi/2}$ rotates by $\pi/2$;

– multiplication by $-i = e^{-i\pi/2}$ rotates by $-\pi/2$;

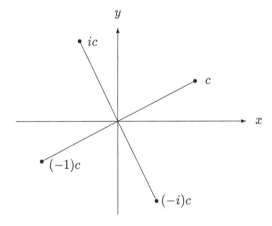

Example 2.6

Find the real and imaginary parts of $c = 1/(1 + e^{i\theta})$.

Solution

One way is to use the standard method of multiplying the denominator by its conjugate, obtaining

$$c = \frac{1 + e^{-i\theta}}{(1 + e^{i\theta})(1 + e^{-i\theta})} = \frac{(1 + \cos\theta) - i\sin\theta}{2 + 2\cos\theta},$$

and hence

$$\operatorname{Re} c = \frac{1}{2}, \quad \operatorname{Im} c = \frac{-\sin\theta}{2 + 2\cos\theta}.$$

More ingeniously, we can multiply the numerator and denominator by $e^{-i(\theta/2)}$, obtaining

$$c = \frac{e^{-i(\theta/2)}}{e^{-i(\theta/2)} + e^{i(\theta/2)}} = \frac{\cos(\theta/2) - i\sin(\theta/2)}{2\cos(\theta/2)},$$

and hence

$$\operatorname{Re} c = \frac{1}{2}, \quad \operatorname{Im} c = -\frac{1}{2}\tan(\theta/2).$$

The verification that the two answers for the imaginary part are actually the same is a simple trigonometrical exercise.   □

Example 2.7

Sum the (finite) series

$$C = 1 + \cos\theta + \cos 2\theta + \cdots + \cos n\theta,$$

where θ is not an integral multiple of 2π.

Solution

Consider the series

$$Z = 1 + e^{i\theta} + e^{2i\theta} + \cdots + e^{ni\theta}.$$

This is a geometric series with common ratio $e^{i\theta}$. The formula for a sum of a geometric series works just as well in $\mathbb{C}$ as in $\mathbb{R}$, and so

$$
\begin{aligned}
Z &= \frac{e^{i(n+1)\theta} - 1}{e^{i\theta} - 1} = \frac{e^{i(n+\frac{1}{2})\theta} - e^{-\frac{1}{2}i\theta}}{e^{\frac{1}{2}i\theta} - e^{-\frac{1}{2}i\theta}} \\
&= \frac{e^{i(n+\frac{1}{2})\theta} - e^{-\frac{1}{2}i\theta}}{2i\sin\frac{1}{2}\theta} \quad \text{(by the Euler formula (2.10))} \\
&= \frac{-i\left(\cos(n+\frac{1}{2})\theta + i\sin(n+\frac{1}{2})\theta\right) + i\left(\cos\frac{1}{2}\theta - i\sin\frac{1}{2}\theta\right)}{2\sin\frac{1}{2}\theta} \quad \text{(since } 1/i = -i) \\
&= \frac{\left(\sin(n+\frac{1}{2})\theta + \sin\frac{1}{2}\theta\right) + i\left(\cos\frac{1}{2}\theta - \cos(n+\frac{1}{2})\theta\right)}{2\sin\frac{1}{2}\theta}.
\end{aligned}
$$

Hence, equating real parts, we deduce that

$$C = \frac{\sin(n + \frac{1}{2})\theta + \sin \frac{1}{2}\theta}{2 \sin \frac{1}{2}\theta} \quad (\theta \neq 2k\pi).$$

As a bonus, our method gives us (if we equate imaginary parts) the result that

$$\sin \theta + \sin 2\theta + \cdots + \sin n\theta = \frac{\cos \frac{1}{2}\theta - \cos(n + \frac{1}{2})\theta}{2 \sin \frac{1}{2}\theta}.$$

$\square$

Example 2.8

Find all the roots of the equation $z^4 + 1 = 0$. Factorise the polynomial in $\mathbb{C}$, and also in $\mathbb{R}$.

Solution

$z^4 = -1 = e^{i\pi}$ if and only if $z = e^{\pm \pi i/4}$ or $e^{\pm 3\pi i/4}$. The roots all lie on the unit circle, and are equally spaced.

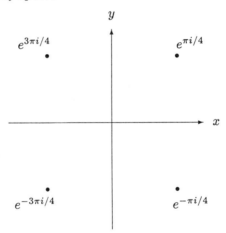

In $\mathbb{C}$ the factorisation is

$$z^4 + 1 = (z - e^{\pi i/4})(z - e^{-\pi i/4})(z - e^{3\pi i/4})(z - e^{-3\pi i/4}).$$

Combining conjugate factors, we obtain the factorisation in $\mathbb{R}$:

$$z^4 + 1 = \left(z^2 - 2z \cos(\pi/4) + 1\right)\left(z^2 - 2z \cos(3\pi/4) + 1\right)$$
$$= (z^2 - z\sqrt{2} + 1)(z^2 + z\sqrt{2} + 1).$$

The strong connections between the operations of complex numbers and the geometry of the plane enable us to specify certain important geometrical objects by means of complex equations. The most obvious case is that of the circle $\{z : |z - c| = r\}$ with centre c and radius $r \geq 0$. This easily translates to the familiar form of the equation of a circle: if $z = x + iy$ and $c = a + ib$, then $|z - c| = r$ if and only if $|z - c|^2 = r^2$, that is, if and only if $(x - a)^2 + (y - b)^2 = r^2$. The other form, $x^2 + y^2 + 2gx + 2fy + c = 0$, of the equation of the circle can be rewritten as $z\bar{z} + hz + \overline{hz} + c = 0$, where $h = g - if$. More generally, we have the equation

$$Az\bar{z} + Bz + \overline{Bz} + C = 0, \qquad (2.11)$$

where $A (\neq 0)$ and C are real, and B is complex. The set

$$\{z \in \mathbb{C} : Az\bar{z} + Bz + \overline{Bz} + C = 0\}$$

is:

(C1) a circle with centre $-\bar{B}/A$ and radius R, where $R^2 = (B\bar{B} - AC)/A^2$ if $B\bar{B} - AC \geq 0$;

(C2) empty if $B\bar{B} - AC < 0$.

If $A = 0$, the equation reduces to

$$Bz + \overline{Bz} + C = 0, \qquad (2.12)$$

and this (provided $B \neq 0$) is the equation of a straight line: if $B = B_1 + iB_2$ and $z = x + iy$ the equation becomes

$$B_1 x - B_2 y + C = 0.$$

Theorem 2.9

Let c, d be distinct complex numbers, and let $k > 0$. Then the set

$$\{z : |z - c| = k|z - d|\}$$

is a circle unless $k = 1$, in which case the set is a straight line, the perpendicular bisector of the line joining c and d.

Proof

We begin with some routine algebra:

$$\begin{aligned}
\{z : |z - c| = k|z - d|\} &= \{z : |z - c|^2 = k^2|z - d|^2\} \\
&= \{z : (z - c)(\bar{z} - \bar{c}) = k^2(z - d)(\bar{z} - \bar{d})\} \\
&= \{z : z\bar{z} - c\bar{z} - \bar{c}z + c\bar{c} = k^2(z\bar{z} - d\bar{z} - \bar{d}z + d\bar{d})\} \\
&= \{z : Az\bar{z} + Bz + \overline{Bz} + C = 0\},
\end{aligned}$$

where $A = k^2 - 1$, $B = \bar{c} - k^2\bar{d}$, $C = k^2 d\bar{d} - c\bar{c}$.

If $k = 1$ we have the set

$$\{z \ : \ (\bar{c} - \bar{d})z + (c - d)\bar{z} + (d\bar{d} - c\bar{c}) = 0\}$$

and this is a straight line. Geometrically, it is clear that it is the perpendicular bisector of the line joining c and d.

If $A \neq 1$ then the set is a circle with centre

$$-\frac{\bar{B}}{A} = \frac{k^2 d - c}{k^2 - 1}, \tag{2.13}$$

for we can show that $B\bar{B} - AC > 0$:

$$\begin{aligned}
B\bar{B} - AC &= (c - k^2 d)(\bar{c} - k^2\bar{d}) - (k^2 - 1)(k^2 d\bar{d} - c\bar{c}) \\
&= c\bar{c} - k^2 c\bar{d} - k^2 \bar{c}d + k^4 d\bar{d} - k^4 d\bar{d} + k^2 d\bar{d} + k^2 c\bar{c} - c\bar{c} \\
&= k^2(c\bar{c} - c\bar{d} - \bar{c}d + d\bar{d}) \\
&= k^2(c - d)(\bar{c} - \bar{d}) = k^2|c - d|^2 > 0.
\end{aligned}$$

The radius of the circle is R, where

$$R^2 = \frac{B\bar{B} - AC}{A^2} = \frac{k^2|c - d|^2}{(k^2 - 1)^2}. \tag{2.14}$$

$\square$

Remark 2.10

The circle $\{z \ : \ |z - c| = k|z - d|\}$ has PQ as diameter. If S is the centre of the circle, then

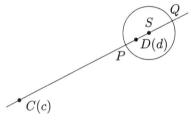

From (2.13) and (2.14) we see that

$$|SC|.|SD| = \left|c - \frac{k^2 d - c}{k^2 - 1}\right| \left|d - \frac{k^2 d - c}{k^2 - 1}\right| = \frac{k^2|c - d|^2}{(k^2 - 1)^2} = R^2. \tag{2.15}$$

We say that the points C and D are **inverse points** with respect to the circle. We shall return to this idea in Chapter 11.

Remark 2.11

The observation that C and D are inverse points is the key to showing that *every* circle can be represented as $\{z : |z - c| = k|z - d|\}$. Suppose that Σ is a circle with centre a and radius R. Let $c = a + t$, where $0 < t < |a|$, and let $d = a + (R^2/t)$. Then c and d are inverse points with respect to Σ. For every point $z = a + Re^{i\theta}$ on Σ,

$$\left|\frac{z - c}{z - d}\right| = \left|\frac{z - c}{\bar{z} - \bar{d}}\right| = \left|\frac{Re^{i\theta} - t}{Re^{-i\theta} - (R^2/t)}\right| = \left|\frac{te^{i\theta}}{R}\right| \left|\frac{R - te^{-i\theta}}{te^{-i\theta} - R}\right| = \frac{t}{R},$$

and so $|z - c| = (t/R)|z - d|$. The answer is not unique.

EXERCISES

2.3. Show that $\mathrm{Re}(iz) = -\mathrm{Im}\, z$, $\mathrm{Im}(iz) = \mathrm{Re}\, z$.

2.4. Write each of the following complex numbers in the standard form $a + bi$, where $a, b \in \mathbb{R}$:

a) $(3 + 2i)/(1 + i)$;

b) $(1 + i)/(3 - i)$;

c) $(z + 2)/(z + 1)$, where $z = x + yi$ with x, y in $\mathbb{R}$.

2.5. Calculate the modulus and principal argument of

$$
\begin{array}{llll}
\text{a)} & 1 - i & \text{b)} & -3i \\
\text{c)} & 3 + 4i & \text{d)} & -1 + 2i
\end{array}
$$

2.6 Show that, for every pair c, d of non-zero complex numbers,

$$\arg(c/d) \equiv \arg c - \arg d \pmod{2\pi}.$$

2.7. Express $1 + i$ in polar form, and hence calculate $(1 + i)^{16}$.

2.8. Show that $(2 + 2i\sqrt{3})^9 = -2^{18}$. (Don't use the binomial theorem!)

2.9. Let $n \in \mathbb{Z}$. Show that, if $n = 4q + r$, with $0 \le r \le 3$, then

$$
i^n = \begin{cases}
1 & \text{if } r = 0 \\
i & \text{if } r = 1 \\
-1 & \text{if } r = 2 \\
-i & \text{if } r = 3.
\end{cases}
$$

2.10. Calculate $\sum_{r=0}^{100} i^r$.

2.11. Show by induction that, for all $z \neq 1$,

$$1 + 2z + 3z^2 + \cdots + nz^{n-1} = \frac{1 - (n+1)z^n + nz^{n+1}}{(1-z)^2}.$$

Deduce that, if $|z| < 1$,

$$\sum_{n=1}^{\infty} nz^{n-1} = \frac{1}{(1-z)^2}.$$

2.12. Let z_1, z_2 be complex numbers such that $|z_1| > |z_2|$. Show that, for all $n \geq 2$,

$$n \left| \frac{z_2}{z_1} \right|^{n-1} < \frac{|z_1|}{|z_1| - |z_2|}.$$

2.13. Prove that, if $z_1, z_2 \in \mathbb{C}$, then

$$|z_1 + z_2|^2 + |z_1 - z_2|^2 = 2(|z_1|^2 + |z_2|^2).$$

Deduce that, for all c, d in $\mathbb{C}$,

$$|c + \sqrt{c^2 - d^2}| + |c - \sqrt{c^2 - d^2}| = |c + d| + |c - d|.$$

2.14. Sum the series

$$\cos\theta + \cos 3\theta + \cdots + \cos(2n+1)\theta.$$

2.15. Let $\gamma = \rho e^{i\theta}$ ($\notin \mathbb{R}$) be a root of $P(z) = 0$, where

$$P(z) = a_n z^n + a_{n-1} z^{n-1} + \cdots + a_1 z + a_0,$$

and where $a_0, a_1, \ldots, a_n$ are *real*. Show that $\bar{\gamma}$ is also a root, and deduce that $z^2 - 2\rho\cos\theta + \rho^2$ is a factor of $P(z)$.

2.16. Determine the roots of the equations

a) $z^2 - (3 - i)z + (4 - 3i) = 0$;

b) $z^2 - (3 + i)z + (2 + i) = 0$.

2.17. Give geometrical descriptions of the sets

a) $\{z : |2z + 3| \leq 1\}$ b) $\{z : |z| \geq |2z + 1|\}$.

2.18. Determine the roots of $z^5 = 1$, and deduce that

$$z^5 - 1 = (z - 1)\left(z^2 - 2z \cos \frac{2\pi}{5} + 1\right)\left(z^2 - 2z \cos \frac{4\pi}{5} + 1\right).$$

Deduce that

$$\cos \frac{2\pi}{5} + \cos \frac{4\pi}{5} = -\frac{1}{2}, \quad \cos \frac{2\pi}{5} \cos \frac{4\pi}{5} = -\frac{1}{4},$$

and hence show that

$$\cos \frac{\pi}{5} = \frac{\sqrt{5} + 1}{4}, \quad \cos \frac{2\pi}{5} = \frac{\sqrt{5} - 1}{4}.$$

3
Prelude to Complex Analysis

3.1 Why is Complex Analysis Possible?

The development of real analysis (sequences, series, continuity, differentiation, integration) depends on a number of properties of the real number system. First, $\mathbb{R}$ is a **field**, a set in which one may add, multiply, subtract and (except by 0) divide. Secondly, there is a notion of **distance**: given two numbers a and b, the distance between a and b is $|a - b|$. Thirdly, to put it very informally, $\mathbb{R}$ has no gaps.

This third property is made more precise by the Least Upper Bound Axiom (see Section 1.2), a property that distinguishes the real number system from the rational number system and is crucial in the development of the theory of real functions. Might something similar hold for complex numbers?

Certainly $\mathbb{C}$ is a field, and the distance between two complex numbers a and b is $|a - b|$. However, the whole idea of a bounded set and a least upper bound depends on the existence of the order relation $\leq$ in $\mathbb{R}$, and there is no useful way of defining a relation $\leq$ in $\mathbb{C}$. In $\mathbb{R}$, the "compatibility" property

$$a \leq b \text{ and } c \geq 0 \quad \Longrightarrow \quad ac \leq bc$$

has the consequence that $a^2 > 0$ for all $a \neq 0$ in $\mathbb{R}$. It is this that makes a useful ordering of $\mathbb{C}$ impossible, for the same compatibility property – and an order without that property would be of no interest – would force us to conclude that

$$1^2 = 1 > 0, \quad i^2 = -1 > 0.$$

Despite the absence of order in $\mathbb{C}$, inequalities play as crucial a role in complex analysis as they do in real analysis. We cannot write a statement $z < w$ concerning complex numbers, but we can, and very frequently do write $|z| < |w|$. (One beneficial effect of the absence of order in $\mathbb{C}$ is that when we say something like "Let $K > 0\ldots$" we do not need to explain that K is real.)

All is not lost, however. In real analysis we can deduce from the Least Upper Bound Axiom the so-called **Completeness Property**. See Theorem 1.3. Informally one can see that this is another version (in fact equivalent to the Least Upper Bound Property) of the "no gaps" property of $\mathbb{R}$. In a Cauchy sequence the terms a_n of the sequence can be made arbitrarily close to each other for sufficiently large n. The property tells us that there is a number, the limit of the sequence, which the terms approach as $n \to \infty$.

It is clear that the definition of a Cauchy sequence and the Completeness Property do make sense if we switch from $\mathbb{R}$ to $\mathbb{C}$. A sequence (c_n) of complex numbers is said to be a **Cauchy sequence** if, for every $\epsilon > 0$ there exists N such that $|a_m - a_n| < \epsilon$ for all $m, n > N$. Then we have:

Theorem 3.1 (Completeness Property of $\mathbb{C}$)

If (c_n) is a Cauchy sequence in $\mathbb{C}$, then (c_n) is convergent.

Proof

We assume the completeness of $\mathbb{R}$. Let $c_n = a_n + ib_n$, where a_n and b_n are real, and suppose that (c_n) is a Cauchy sequence. That is, suppose that for all $\epsilon > 0$ there exists N such that $|c_m - c_n| < \epsilon$ for all $m, n > N$. Now

$$
\begin{aligned}
|c_m - c_n| &= |(a_m - a_n) + i(b_m - b_n)| \\
&= [(a_m - a_n)^2 + (b_m - b_n)^2]^{1/2} \\
&\geq \max\{|a_m - a_n|, |b_m - b_n|\},
\end{aligned}
$$

and so (a_n), (b_n) are both real Cauchy sequences, with limits α, β respectively. Thus, for all $\epsilon > 0$ there exists N_1 such that $|a_n - \alpha| < \epsilon/2$ for all $n > N_1$, and there exists N_2 such that $|b_n - \beta| < \epsilon/2$ for all $n > N_2$. Hence, for all $n > \max\{N_1, N_2\}$,

$$
|c_n - (\alpha + i\beta)| = |(a_n - \alpha) + i(b_n - \beta)| \leq |a_n - \alpha| + |b_n - \beta| < \epsilon,
$$

and so $(c_n) \to \alpha + i\beta$. $\qquad\qquad\square$

As in Section 1.3, the Completeness Property translates for series into

Theorem 3.2 (The General Principle of Convergence)

Let $\sum_{n=1}^{\infty} c_n$ be a series of complex terms. If, for every $\epsilon > 0$, there exists N such that

$$\left| \sum_{r=n+1}^{m} c_r \right| < \epsilon$$

for all $m > n > N$, then $\sum_{n=1}^{\infty} c_n$ is convergent.

In general terms, because the definitions involve inequalities of the type $|a| < |b|$, the notions of limit and convergence for sequences and series apply to the complex case without alteration. The geometry of the plane is, however, more complicated than that of the line, and the next section draws attention to this aspect.

3.2 Some Useful Terminology

In real analysis one makes frequent reference to **intervals**, and notations such as $[a, b)$ for the set $\{x \in \mathbb{R} : a \leq x < b\}$ are very useful. In complex analysis the situation is inevitably more complicated, since we are usually dealing with subsets of the plane rather than of the line. It is therefore convenient at this point to introduce some ideas and terms that will make our statements less cumbersome.

First, if $c \in \mathbb{C}$ and $r > 0$, we shall denote the set $\{z \in \mathbb{C} : |z - c| < r\}$ by $N(c, r)$. We shall call it a **neighbourhood** of c, or, if we need to be more specific, the r-**neighbourhood** of c.

Next, a subset U of $\mathbb{C}$ will be called **open** if, for all u in U, there exists $\delta > 0$ such that $N(u, \delta) \subset U$.

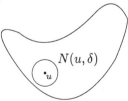

Among important cases of open sets, apart from $\mathbb{C}$ itself, are

$$\mathbb{C} \setminus \{c\}, \quad N(c, r), \quad \{z \in \mathbb{C} : |z - c| > r\}.$$

The empty set $\emptyset$ is also open; the definition applies "vacuously", for in this case there are no elements u in U.

A subset D of $\mathbb{C}$ will be called **closed** if its complement $\mathbb{C} \setminus D$ in $\mathbb{C}$ is open, that is, if, for all $z \notin D$ there exist $\delta > 0$ such that $N(z, \delta)$ lies wholly outside D. Among closed subsets of $\mathbb{C}$ are $\mathbb{C}$, $\emptyset$, any finite subset of $\mathbb{C}$, $\{z \in \mathbb{C} : |z - c| \le r\}$, $\{z \in \mathbb{C} : |z - c| \ge r\}$. There exist sets that are neither open nor closed (see Example 3.4), but the only two subsets of $\mathbb{C}$ that are both open and closed are $\mathbb{C}$ and $\emptyset$. (See Example 3.5 below.)

The **closure** $\overline{S}$ of a subset S of $\mathbb{C}$ is defined as the set of elements z with the property that *every* neighbourhood $N(z, \delta)$ of z has a non-empty intersection with S. The **interior** $I(S)$ of S is the set of z in S for which some neighbourhood $N(z, \delta)$ of z is wholly contained in S. The **boundary** ∂S of S is defined as $\overline{S} \setminus I(S)$.

Theorem 3.3

Let S be a non-empty subset of $\mathbb{C}$.

(i) $I(S)$ is open.   (ii) $S = I(S)$ if and only if S is open.

(iii) $\overline{S}$ is closed.   (iv) $S = \overline{S}$ if and only if S is closed.

Proof*

(i) If $I(S) = \emptyset$ the result is clear. Otherwise, let $z \in I(S)$. By definition, there exists a neighbourhood $N(z, \delta)$ of z wholly contained in S. Let $w \in N(z, \delta)$. Since $N(z, \delta)$ is open, there exists a neighbourhood $N(w, \epsilon)$ of w such that

$$N(w, \epsilon) \subset N(z, \delta) \subset S .$$

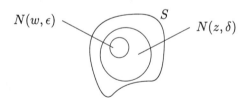

Thus $w \in I(S)$, and, since w was an arbitrary element of $N(z, \delta)$, we conclude that $N(z, \delta) \subset I(S)$. Thus $I(S)$ is open.

(ii) From Part (i), $S = I(S)$ immediately implies that S is open. For the converse, suppose that S is open, and let $z \in S$. Then there exists $N(z, \delta)$ wholly contained in S, and so, by definition, $z \in I(S)$. Since it is clear that $I(S) \subseteq S$, we deduce that $S = I(S)$.

(iii) If $\mathbb{C}\setminus\overline{S}$ is empty then it is certainly open, and so $\overline{S} = \mathbb{C}$ is closed. Otherwise, let $z \in \mathbb{C}\setminus\overline{S}$. Then it is *not* the case that every $N(z,\delta)$ intersects S, and so there exists some $N(z,\delta)$ wholly contained in $\mathbb{C}\setminus S$. Let $w \in N(z,\delta)$. As in Part (i) we must have a neighbourhood $N(w,\epsilon)$ of w such that

$$N(w,\epsilon) \subset N(z,\delta) \subset \mathbb{C}\setminus S\,.$$

Thus $w \notin \overline{S}$ and, since this holds for every w in $N(z,\delta)$, we deduce that $N(z,\delta) \subset \mathbb{C}\setminus\overline{S}$. Thus $\mathbb{C}\setminus\overline{S}$ is open, and hence $\overline{S}$ is closed.

(iv) From Part (iii), $S = \overline{S}$ immediately implies that S is closed. It is clear from the definition that $S \subseteq \overline{S}$ for every set S. Suppose now that S is closed, and let $z \in \overline{S}$. Then every $N(z,\delta)$ intersects S. If $z \notin S$, then, since $\mathbb{C}\setminus S$ is open, there exists $N(z,\epsilon)$ wholly contained in $\mathbb{C}\setminus S$, and we have a contradiction. Hence $z \in S$, and so we have proved that $S = \overline{S}$.

$\square$

Example 3.4

Let

$$S = \{(x,y) \,:\, -a < x < a,\ -b \le y \le b\}\,.$$

Find $\mathrm{I}(S)$, $\overline{S}$, ∂S, and deduce that S is neither open nor closed.

Solution

$$\mathrm{I}(S) = \{(x,y) \,:\, -a < x < a,\ -b < y < b\}\,,$$
$$\overline{S} = \{(x,y) \,:\, -a \le x \le a,\ -b \le y \le b\}\,,$$
$$\partial S = T \cup U \cup V \cup W\,,$$

where

$$T = \{(a,y) \,:\, -b \le y \le b\}\,, \quad U = \{(-a,y) \,:\, -b \le y \le b\}\,,$$
$$V = \{(x,b) \,:\, -a \le x \le a\}\,, \quad W = \{(x,-b) \,:\, -a \le x \le a\}\,.$$

Since $S \ne \mathrm{I}(S)$ and $S \ne \overline{S}$, it follows from Theorem 3.3 that S is neither open nor closed.

$\square$

Example 3.5

Show that $\mathbb{C}$ and $\emptyset$ are the only two subsets of $\mathbb{C}$ that are both open and closed.

Solution

Let U be both open and closed. Suppose, for a contradiction, that U and $\mathbb{C}\setminus U$ are both non-empty, and let $z \in U$, $w \in \mathbb{C}\setminus U$.

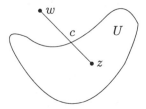

Let

$$\sigma = \sup\left\{\lambda \in (0,1) \ : \ (1-\lambda)z + \lambda w \in U\right\}.$$

Let $c = (1 - \sigma)z + \sigma w$. Then either $c \in U$ or $c \in \mathbb{C}\setminus U$. Suppose first that $c \in U$. Since U is open, there is a neighbourhood $N(c, \delta)$ lying wholly within U, and so in particular there is a $\lambda > \sigma$ for which $(1 - \lambda)z + \lambda w \in U$. This is a contradiction. On the other hand, suppose that $c \in \mathbb{C}\setminus U$. Then, since $\mathbb{C}\setminus U$ is open, there is a neighbourhood $N(c, \delta)$ lying wholly within $\mathbb{C}\setminus U$. Thus there exists $\tau < \sigma$ such that τ is an upper bound of $\{\lambda \in (0,1) \ : \ (1 - \lambda)z + \lambda w \in U\}$. This too is a contradiction. Hence $\mathbb{C}$ and $\emptyset$ are the only open and closed sets in $\mathbb{C}$.

We shall use the following notations and terminology:

- $N(a, r) = \{z \ : \ |z - a| < r\}$, a **neighbourhood**, an **open disc**;
- $\overline{N}(a, r) = \{z \ : \ |z - a| \leq r\}$, a **closed disc**, the closure of $N(a, r)$;
- $\kappa(a, r) = \{z \ : \ |z - a| = r\} = \partial N(a, r) = \partial\overline{N}(a, r)$, a **circle**;
- $D'(a, r) = \{z \ : \ 0 < |z - a| < r\} = N(a, r) \setminus \{0\}$, a **punctured disc**.

EXERCISES

3.1. Show that a closed interval $[a, b]$ on the real line is a closed subset of $\mathbb{C}$, but that an open interval (a, b) is *not* an open subset of $\mathbb{C}$. Is it closed?

3.2. Show that the set $A = \{z \in \mathbb{C} : 1 < |z| < 2\}$ is open. Describe its closure and its boundary.

3.3 Functions and Continuity

In complex analysis, as in other areas of mathematics, we think of a **function** as a "process" transforming one complex number into another. While this process may involve a complicated description, most of the important cases involve the use of a formula. Frequently the function is defined only for a subset of $\mathbb{C}$, and we talk of the **domain of definition**, or simply the **domain**, of the function. Thus the function $z \mapsto 1/z$ has domain of definition $\mathbb{C} \setminus \{0\}$.

If $z = x + iy \in \mathbb{C}$ and $f : \mathbb{C} \to \mathbb{C}$ is a function, then there are real functions u and v of two variables such that

$$f(z) = u(x, y) + iv(x, y). \tag{3.1}$$

For example, if $f(z) = z^2$, then

$$f(z) = (x + iy)^2 = (x^2 - y^2) + i(2xy), \tag{3.2}$$

and so

$$u(x, y) = x^2 - y^2, \quad v(x, y) = 2xy.$$

We refer to u and v as the **real** and **imaginary** parts of f, and write $u = \operatorname{Re} f$, $v = \operatorname{Im} f$. By $|f|$ we shall mean the function $z \mapsto |f(z)|$.

We cannot draw graphs of complex functions in the way that we do for real functions, since the graph $\{(z, f(z)) : z \in \mathbb{C}\}$ would require four dimensions. What can be useful is to picture z and $w = f(z)$ in two complex planes, and it can be instructive to picture the image in the w-plane of a path in the z-plane.

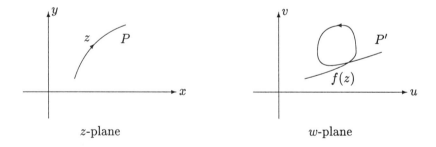

z-plane w-plane

As the point z moves along the path P in the z-plane, its image $f(z)$ traces out the path P' in the w-plane.

For the function $z \mapsto z^2$, the hyperbolic curves $x^2 - y^2 = k$ and $2xy = l$ in the z-plane transform (see Figure 3.1) respectively to the straight lines $u = k$ and $v = l$ in the w-plane.

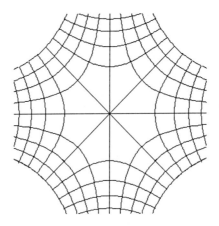

Figure 3.1. The z-plane

Less obviously, but by a routine calculation, we see that the straight lines $x = k$ and $y = l$ in the z-plane transform respectively (see Figure 3.2) to the parabolic curves

$$v^2 = 4k^2(k^2 - u), \quad v^2 = 4l^2(u + l^2)$$

in the w-plane:

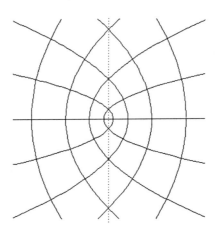

Figure 3.2. The w-plane

The concept of a **limit** is, as in real analysis, central to the development of our subject. Given a complex function f and complex numbers l and c, we say that $\lim_{z \to c} f(z) = l$ if, for every $\epsilon > 0$, there exists $\delta > 0$ such that $|f(z) - l| < \epsilon$ for all z such that $0 < |z - c| < \delta$. Formally this definition is exactly the same as for real functions, and many real analysis proofs can simply be reinterpreted as proofs in complex analysis. Our accompanying picture is different, however, for the set $\{z \in \mathbb{C} : 0 < |z - c| < \delta\}$ is not a punctured interval, but a punctured disc:

The two-dimensional nature of the definition does introduce some complications. For real numbers there are in effect only two directions along which a sequence of points can approach a real number c, but in the complex plane there are infinitely many paths towards c, and the existence of a limit requires that the limit should exist for every possible path.

Example 3.6

Let $f(x + iy) = u(x, y) + iv(x, y)$, where $v(x, y) = 0$ for all x, y, and

$$u(x, y) = \frac{xy}{x^2 + y^2} \quad ((x, y) \neq (0, 0)) .$$

Show that $\lim_{x \to 0} f(x+i0)$ and $\lim_{y \to 0} f(0+iy)$ both exist, but that $\lim_{z \to 0} f(z)$ does not.

Solution

Since $u(x, 0) = 0$ for all x and $u(0, y) = 0$ for all y, it is clear that

$$\lim_{x \to 0} f(x + i0) = \lim_{y \to 0} f(0 + iy) = 0 .$$

On the other hand, if $z = re^{i\theta} = r(\cos\theta + i\sin\theta)$,

$$|f(z)| = \left| \frac{r^2 \cos\theta \sin\theta}{r^2} \right| = |\cos\theta \sin\theta| .$$

If, for example, $\theta = \pi/4$, then $|f(z)| = 1/2$ for all values of r. Thus, no matter how small ϵ may be, within the neighbourhood $N(0, \epsilon)$ there is a $z \ (= (\epsilon/2)+i0)$ for which $f(z) = 0$ and a $z \ (= (\epsilon/2)e^{i\pi/4})$ for which $f(z) = 1/2$. The limit does not exist.   □

In that example we obtained differing limiting values when 0 was approached from different angles by straight line paths. It might be thought that the limit of $f(z)$ as $z \to 0$ would exist if all these straight line paths led to the same result. But even this is not so:

Example 3.7

Let $f(x+iy) = u(x,y) + iv(x,y)$, where $v(x,y) = 0$ for all x, y, and

$$u(x,y) = \frac{xy^2}{x^2 + y^4} \quad ((x,y) \neq (0,0)) \,.$$

Show that, for all θ, $\lim_{r\to 0} f(re^{i\theta}) = 0$, but that $\lim_{z\to 0} f(z)$ does not exist.

Solution

$$|f(re^{i\theta})| = \left| \frac{r^3 \cos\theta \sin^2\theta}{r^2 \cos^2\theta + r^4 \sin^4\theta} \right| \,.$$

If $\cos\theta = 0$ then $|f(re^{i\theta})| = 0$ for all r. Otherwise, for any fixed θ such that $\cos\theta \neq 0$,

$$|f(re^{i\theta})| \leq \left| \frac{r^3}{r^2 \cos^2\theta + r^4 \sin^4\theta} \right| \leq \left| \frac{r^3}{r^2 \cos^2\theta} \right| = \frac{r}{\cos^2\theta} < \epsilon$$

for sufficiently small r, and so $\lim_{r\to 0} f(re^{i\theta}) = 0$. On the other hand,

$$f(y^2 + yi) = \frac{y^4}{y^4 + y^4} = \frac{1}{2} \,,$$

and so there are points z arbitrarily close to 0 for which $f(z) = 1/2$. The limit does not exist.   □

Naturally, we are chiefly interested in cases where the limit *does* exist, and especially in continuous functions: a complex function f is said to be **continuous** at a point c if $\lim_{z\to c} f(z) = f(c)$. To spell it out in full, f is continuous at c if, for all $\epsilon > 0$, there exists $\delta > 0$ such that $|f(z) - f(c)| < \epsilon$ for all z in the punctured disc $D'(c,\delta)$.

Example 3.8

Show that the function $f : z \mapsto |z|^2$ is continuous at every point c.

Solution

Let $\epsilon > 0$ be given. Observe that

$$|f(z) - f(c)| = \left| |z|^2 - |c|^2 \right| = \left| |z| - |c| \right| (|z| + |c|) \leq |z - c| (|z| + |c|) \,.$$

Let $\delta \leq 1$. Then $0 < |z - c| < 1$ implies $|z| - |c| < 1$ (by Theorem 2.1), and so $|z| < |c| + 1$. Hence

$$|f(z) - f(c)| \leq (2|c| + 1) |z - c| \,.$$

Hence, if $\delta = \min\{1, \epsilon/(2|c|+1)\}$, then $z \in D'(c,\delta)$ implies that $|f(z) - f(c)| < \epsilon$. Thus f is continuous at c.   □

Remark 3.9

The standard "calculus of limits", familiar for real functions, applies also to complex functions, and the proofs are formally identical. (See [9, Chapter 3].) If $\lim_{z \to c} f(z) = l$ and $\lim_{z \to c} g(z) = m$, then $kf(z)$, $f(z) \pm g(z)$, $f(z)g(z)$ and (provided $m \neq 0$) $f(z)/g(z)$ have limits kl, $l \pm m$, lm and l/m, respectively. Also, the continuity of f and g at c implies the continuity of kf, $f \pm g$, $f \cdot g$ and (unless $g(c) = 0$) f/g.

We shall also be interested in limits as $z \to \infty$, and here there is a potential difficulty, since there are many paths to infinity on the complex plane. The obvious definition is that $\lim_{z \to \infty} f(z) = L$ if, for every $\epsilon > 0$, there exists $K > 0$ such that $|f(z) - L| < \epsilon$ whenever $|z| > K$. Similarly, $f(z) \to \infty$ as $z \to \infty$ if, for all $E > 0$ there exists $D > 0$ such that $|f(z)| > E$ whenever $|z| > D$. It can sometimes be useful to think of ∞ as a single point, and to extend the complex plane by adjoining that point. If this seems artificial, one can change the visualisation of complex numbers by thinking of the complex plane as the equatorial plane of a sphere of radius 1, with north pole N.

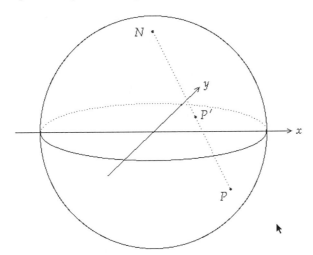

For each $P \neq N$ on the sphere we define P' as the point in which NP meets the equatorial plane. This gives a one-to-one correspondence between the points P on the sphere (except N) and the points on the plane. We may thus visualise the complex numbers as points on a sphere (the **Riemann**[1] **sphere**, and the "missing" point N is the **point at infinity**. For the most part, however, it is sufficient to note that $\lim_{z \to \infty} f(z)$ is the same as the more easily understood $\lim_{|z| \to \infty} f(z)$.

---

[1] Georg Friedrich Bernhard Riemann, 1826–1866.

3.4 The O and o Notations

This is a convenient place to introduce some extremely useful notations which enable us to grasp the essence of analytic arguments without unnecessary detail. In analysis one frequently requires to prove that a quantity q is "small". That is an over-simplification, for what is small depends on the context: we do not normally think of the diameter of the earth's orbit as small, but it is indeed small compared with the distance to the edge of the galaxy. So, more precisely, one requires in fact to prove that q is small by comparison with another quantity Q. Furthermore, we are always interested in the *ultimate* comparison between the quantities, and so to say that q is small compared with Q is to say that, in the limit, the ratio q/Q is zero. We can save a lot of unnecessary technicalities by using O and o notations, which we now proceed to explain.

Let f, ϕ be complex functions (and here there is a tacit assumption that ϕ is in some way "better known" than f); then

- $f(z) = O(\phi(z))$ as $z \to \infty$ means that there is a positive constant K such that $|f(z)| \le K|\phi(z)|$ for all sufficiently large $|z|$;

- $f(z) = O(\phi(z))$ as $z \to 0$ means that there is a positive constant K such that $|f(z)| \le K|\phi(z)|$ for all sufficiently small $|z|$;

- $f(z) = o(\phi(z))$ as $z \to \infty$ means that $\lim_{|z|\to\infty} f(z)/\phi(z) = 0$;

- $f(z) = o(\phi(z))$ as $z \to 0$ means that $\lim_{z\to 0} f(z)/\phi(z) = 0$.

We can use the notation in a very flexible way, writing, for example, $O(\phi)$ for any function f with the property that $|f(z)| \le K|\phi(z)|$ for sufficiently small (or sufficiently large) z.

Example 3.10

Show that, as $z \to 0$,

1. $O(z) + O(z) = O(z)$, $o(z) + o(z) = o(z)$;

2. for all $K \ne 0$, $K\,O(z) = O(z)$, $K\,o(z) = o(z)$;

3. $O(z)O(z) = O(z^2)$, $O(z)O(z) = o(z)$.

Solution

1. Let $f_1(z) = O(z)$, $f_2(z) = O(z)$. Thus there exist positive constants K_1, K_2 such that, for all sufficiently small z,

$$|f_1(z)| \le K_1|z|\,, \quad |f_2(z)| \le K_2|z|\,.$$

Hence
$$|f_1(z) + f_2(z)| \leq |f_1(z)| + |f_2(z)| \leq (K_1 + K_2)|z|,$$
and so $f_1(z) + f_2(z) = O(z)$.

Suppose now that $f_1(z) = o(z)$, $f_2(z) = o(z)$. Then, as $z \to 0$, $f_1(z)/z \to 0$ and $f_1(z)/z \to 0$. It follows immediately that
$$\frac{f_1(z) + f_2(z)}{z} = \frac{f(z)}{z} + \frac{f_2(z)}{z} \to 0$$
and so $f_1(z) + f_2(z) = o(z)$.

2. Let $f(z) = O(z)$, so that there exists a positive M such that $|f(z)| \leq M|z|$ for all sufficiently small z. Hence
$$|Kf(z)| = |K||f(z)| \leq M|K||z|,$$
and so $Kf(z) = O(z)$.

Let $f(z) = o(z)$, so that $f(z)/z \to 0$ as $z \to 0$. Hence $Kf(z)/z \to 0$, and so $Kf(z) = o(z)$.

3. Let $f_1(z) = O(z)$, $f_2(z) = O(z)$, so that there are constants K_1, K_2 such that, for sufficiently small z,
$$|f_1(z)| \leq K_1|z|, \quad |f_2(z)| \leq K_2|z|.$$
Then $|f_1(z)f_2(z)| \leq K_1 K_2|z^2|$, and so $f_1(z)f_2(z) = O(z^2)$. Also,
$$\left| \frac{f_1(z)f_2(z)}{z} \right| \leq K_1 K_2|z| \to 0,$$
and so $f_1(z)f_2(z) = o(z)$.

$\square$

Note that $f(z) = o(\phi(z))$ implies that $f(z) = O(\phi(z))$, but that the converse is not true: $1 + z = O(1)$ as $z \to 0$, but it is not true that $1 + z = o(1)$.

Example 3.11

Show that, as $z \to 0$,
$$\frac{1}{(1+z)^2} = 1 - 2z + O(z^2).$$

Solution

By an easy calculation, for all $|z| \leq \frac{1}{2}$,

$$
\left| \frac{1}{(1+z)^2} - (1-2z) \right| = \left| \frac{1 - (1-2z)(1+2z+z^2)}{(1+z)^2} \right|
$$

$$
= \frac{|3z^2 + 2z^3|}{|1+z|^2} \leq \frac{3|z|^2 + 2|z|^3}{(1-|z|)^2}
$$

$$
\leq \frac{4|z|^2}{1/4} = 16|z|^2,
$$

since $|z| \leq \frac{1}{2}$ implies that $(1-|z|)^2 \geq \frac{1}{4}$ and $3|z|^2 + 2|z|^3 \leq 4|z|^2$. Thus

$$
\frac{1}{(1+z)^2} - (1-2z) = O(z^2),
$$

and the result follows.   □

Example 3.12

Show that, as $z \to \infty$,

$$
\frac{4z^3 + 7}{(z^2+2)^2} = \frac{4}{z} - \frac{16}{z^3} + O\left(\frac{1}{z^4}\right).
$$

Solution

$$
\frac{4z^3 + 7}{(z^2+2)^2} = \frac{4 + (7/z^3)}{z\left(1 + (2/z^2)\right)^2}
$$

$$
= \frac{1}{z}\left(4 + \frac{7}{z^3}\right)\left[1 - \frac{4}{z^2} + O\left(\frac{1}{z^4}\right)\right] \quad \text{(by Example 3.11)}
$$

$$
= \frac{4}{z} - \frac{16}{z^3} + O\left(\frac{1}{z^4}\right).
$$

EXERCISES

3.3. Show that

$$
O(z^2) + O(z^3) = \begin{cases} O(z^2) & \text{as } z \to 0 \\ O(z^3) & \text{as } z \to \infty. \end{cases}
$$

3.4 Show that, for all positive integers n,

$$
(1+z)^n = 1 + nz + o(z)
$$

as $z \to 0$.

3.5. Show that, as $z \to \infty$,

$$\frac{3z^2 + 7z + 5}{(z+1)^2} = 3 + \frac{1}{z} + O(z^{-2}).$$

3.6 Let $p(z) = a_0 + a_1 z + \cdots + a_n z^n$, where $n \geq 1$ and $a_n \neq 0$. Show that

$$p(z) = O(1) \text{ as } z \to 0, \qquad p(z) = O(z^n) \text{ as } z \to \infty.$$

4

Differentiation

4.1 Differentiability

The definition of differentiability of a complex function presents no problem, since it is essentially the same as for a real function: a complex function f is said to be **differentiable** at a point c in $\mathbb{C}$ if

$$\lim_{z \to c} \frac{f(z) - f(c)}{z - c}$$

exists. The limit is called the **derivative of f at** c and is denoted by $f'(c)$. Although the definition is formally identical to that used in real analysis, the fact that differentiability requires the rate of change of f to be the same in all possible directions means that its consequences are much more far-reaching. Certain things, however, do not change: the standard "calculus" rules for differentiation of sums, products and quotients, and the "chain rule" $(f \circ g)'(z) = f'(g(z))g'(z)$ are all valid for complex functions, and the proofs are in essence formally identical to those in real analysis. (See [9, Chapter 4].) Since (trivially) $z \mapsto z$ is differentiable, with derivative 1, it follows that polynomials are differentiable at every point in the plane, and that a rational function $p(z)/q(z)$ (where p and q are polynomials with no common factor) is differentiable except at the zeros of q.

If, as usual, we write $f(x + iy)$ as $u(x, y) + iv(x, y)$, and write c as $a + ib$, then, keeping b fixed, we have

$$\frac{f(x + ib) - f(a + ib)}{(x + ib) - (a + ib)} = \frac{u(x, b) - u(a, b)}{x - a} + i \frac{v(x, b) - v(a, b)}{x - a} .$$

Hence the existence of $f'(c)$ implies the existence of the limits

$$\lim_{x \to a} \frac{u(x,b) - u(a,b)}{x - a} \quad \text{and} \quad \lim_{x \to a} \frac{v(x,b) - v(a,b)}{x - a},$$

that is, the existence at the point (a,b) of the partial derivatives $\partial u / \partial x$ and $\partial v / \partial x$. Moreover

$$f'(c) = \frac{\partial u}{\partial x} + i\frac{\partial v}{\partial x}. \tag{4.1}$$

If we now keep a fixed, we see that

$$\begin{aligned}
\frac{f(a + iy) - f(a + ib)}{(a + iy) - (a + ib)} &= \frac{u(a,y) - u(a,b)}{i(y - b)} + \frac{i\big(v(a,y) - v(a,b)\big)}{i(y - b)} \\
&= \frac{v(a,y) - v(a,b)}{(y - b)} - i\frac{u(a,y) - u(a,b)}{y - b},
\end{aligned}$$

and so the differentiability of f at c implies the existence at (a,b) of the partial derivatives $\partial u / \partial y$ and $\partial v / \partial y$. Moreover,

$$f'(c) = \frac{\partial v}{\partial y} - i\frac{\partial u}{\partial y}. \tag{4.2}$$

Comparing (4.1) and (4.2) gives the **Cauchy–Riemann equations**

$$\frac{\partial u}{\partial x} = \frac{\partial v}{\partial y}, \quad \frac{\partial v}{\partial x} = -\frac{\partial u}{\partial y}. \tag{4.3}$$

We record our observations thus far:

Theorem 4.1

Let f be a complex function, differentiable at $c = a + ib$, and suppose that $f(x + iy) = u(x,y) + iv(x,y)$, where x, y, $u(x,y)$ and $v(x,y)$ are real. Then the partial derivatives $\partial u / \partial x$, $\partial v / \partial x$, $\partial u / \partial y$ and $\partial v / \partial y$ all exist at the point (a,b), and

$$\frac{\partial u}{\partial x} = \frac{\partial v}{\partial y}, \quad \frac{\partial v}{\partial x} = -\frac{\partial u}{\partial y}.$$

$\square$

We thus have a necessary condition for differentiability at c. Is it also sufficient? Well, not quite. Consider the following example:

Example 4.2

Let $f(x + iy) = u(x, y) + iv(x, y)$, where

$$u(x, y) = \sqrt{|xy|}, \quad v(x, y) = 0.$$

Show that the Cauchy–Riemann equations are satisfied at $z = 0$, but that f is not differentiable at that point.

Solution

It is clear that $\partial v/\partial x = \partial v/\partial y = 0$, and almost as clear that $\partial u/\partial x = \partial u/\partial y = 0$ at the point $(0, 0)$, since the function $u(x, y)$ takes the constant value 0 along both the x- and the y-axes. More formally, at the point $(0, 0)$,

$$\frac{\partial u}{\partial x} = \lim_{x \to 0} \frac{u(x, 0) - u(0, 0)}{x - 0} = \lim_{x \to 0} \frac{0}{x} = 0,$$

and the computation for $\partial u/\partial y$ is essentially identical. Thus the Cauchy–Riemann equations are trivially satisfied.

On the other hand,

$$\frac{f(z) - f(0)}{z - 0} = \frac{\sqrt{|xy|}}{x + iy} = \frac{\sqrt{|\cos\theta \sin\theta|}}{\cos\theta + i\sin\theta} = \sqrt{|\cos\theta \sin\theta|}\, e^{-i\theta},$$

where $x = r\cos\theta$ and $y = r\sin\theta$. The expression on the right is independent of r, and so $\big(f(z) - f(0)\big)/(z - 0)$ takes this constant value for points on the line $x\sin\theta - y\cos\theta = 0$ arbitrarily close to 0. For $\theta = 0$ or $\theta = \pi/2$ the constant value is 0, but (for example) for $\theta = \pi/4$ the value is $(1 - i)/2$. We are forced to conclude that the limit does not exist.

$\square$

The Cauchy–Riemann equations arise from the requirement that the rate of change of the function at the point c must be the same in the x- and y-directions, and with hindsighht it was probably unreasonable to suppose that they might be a sufficient condition for differentiability, for there are many other ways of approaching the point c. Remarkably, they do come close:

Theorem 4.3

Let $D = N(c, R)$ be an open disc in $\mathbb{C}$. Let f be a complex function whose domain contains D, let $f(x + iy) = u(x, y) + iv(x, y)$, and suppose that the partial derivatives $\partial u/\partial x$, $\partial v/\partial x$, $\partial u/\partial y$, $\partial v/\partial y$ exist and are continuous throughout D. Suppose also that the Cauchy–Riemann equations are satisfied at c. Then f is differentiable at c.

Proof*

Let $c = a + ib$ and let $l = h + ik$ be such that the neighbourhood $N(c, |l|)$ lies entirely in D.

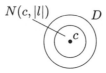

By a standard argument in two-variable calculus,

$$u(a + h, b + k) - u(a, b) = [u(a + h, b + k) - u(a, b + k)]$$
$$+ [u(a, b + k) - u(a, b)]$$
$$= h \frac{\partial u}{\partial x}(a + \theta h, b + k) + k \frac{\partial u}{\partial y}(a, b + \phi k) ,$$

by the mean value theorem, where $\theta, \phi \in (0, 1)$. Since the partial derivatives are continuous, we may deduce that

$$u(a + h, b + k) - u(a, b) = h \left(\frac{\partial u}{\partial x} + \epsilon_1 \right) + k \left(\frac{\partial u}{\partial y} + \epsilon_2 \right)$$

where $\epsilon_1, \epsilon_2 \to 0$ as $h, k \to 0$, and where the partial derivatives are now evaluated at (a, b). Similarly,

$$v(a + h, b + k) - v(a, b) = h \left(\frac{\partial v}{\partial x} + \epsilon_3 \right) + k \left(\frac{\partial v}{\partial y} + \epsilon_4 \right)$$

where again $\epsilon_3, \epsilon_4 \to 0$ as $h, k \to 0$. Hence

$$f(c + l) - f(c) = h \left(\frac{\partial u}{\partial x} + \epsilon_1 \right) + k \left(\frac{\partial u}{\partial y} + \epsilon_2 \right) + ih \left(\frac{\partial v}{\partial x} + \epsilon_3 \right) + ik \left(\frac{\partial v}{\partial y} + \epsilon_4 \right)$$
$$= (h + ik) \left(\frac{\partial u}{\partial x} + i \frac{\partial v}{\partial x} \right) + \epsilon_1 h + \epsilon_2 k + i\epsilon_3 h + i\epsilon_4 k ,$$

by the Cauchy–Riemann equations. Thus

$$\frac{f(c + l) - f(c)}{l} = \frac{\partial u}{\partial x} + i \frac{\partial v}{\partial x} + \frac{1}{l}(\epsilon_1 h + \epsilon_2 k + i\epsilon_3 h + i\epsilon_4 k) \to \frac{\partial u}{\partial x} + i \frac{\partial v}{\partial x}$$

as $l \to 0$. Thus $f'(c)$ exists, and equals $\partial u / \partial x + i(\partial v / \partial x)$. □

A slight modification of this argument, in which c is replaced by an arbitrary point w in D, gives:

Theorem 4.4

Let D be an open disc in $\mathbb{C}$. Let f be a complex function whose domain contains D, let $f(x + iy) = u(x,y) + iv(x,y)$, and suppose that the partial derivatives $\partial u/\partial x$, $\partial v/\partial x$, $\partial u/\partial y$, $\partial v/\partial y$ exist and are continuous throughout D. Suppose also that the Cauchy–Riemann equations are satisfied at all points in D. Then f is differentiable at all points in D. $\qquad\qquad\square$

Remark 4.5

What goes wrong in Example 4.2 is that the partial derivatives are not continuous at the origin. If $x, y > 0$, then

$$\frac{\partial f}{\partial x} = \frac{1}{2}\sqrt{\frac{y}{x}}\,, \quad \frac{\partial f}{\partial y} = \frac{1}{2}\sqrt{\frac{x}{y}}\,,$$

and neither of these has a limit as $(x,y) \to (0,0)$.

Let U be an open set in $\mathbb{C}$. A complex function is said to be **holomorphic in** U if it is differentiable at every point in U. A function which is differentiable at every point in $\mathbb{C}$ is called an **entire** function. From our opening remarks about differentiability, it is clear that every polynomial is an entire function. A rational function such as $z \mapsto (z+i)/(z+1)$ is holomorphic in the set $\mathbb{C}\setminus\{-1\}$.

Some examples are instructive:

Example 4.6

Verify the Cauchy–Riemann equations for the function $z \mapsto z^2$.

Solution

Here

$$u(x,y) = x^2 - y^2\,, \quad v(x,y) = 2xy\,,$$

and so

$$\frac{\partial u}{\partial x} = \frac{\partial v}{\partial y} = 2x\,, \quad \frac{\partial v}{\partial x} = -\frac{\partial u}{\partial y} = 2y\,.$$

The partial derivatives are continuous and the Cauchy–Riemann equations are satisfied. Observe also that the derivative is

$$\frac{\partial u}{\partial x} + i\frac{\partial v}{\partial x} = 2(x + iy) = 2z\,.$$

$\qquad\qquad\square$

Example 4.7

Verify the Cauchy–Riemann equations for the function $z \mapsto 1/z$ $(z \neq 0)$.

Solution

Here

$$u(x, y) = \frac{x}{x^2 + y^2}, \qquad v(x, y) = \frac{-y}{x^2 + y^2},$$

and

$$\frac{\partial u}{\partial x} = \frac{\partial v}{\partial y} = \frac{y^2 - x^2}{(x^2 + y^2)^2}, \qquad \frac{\partial v}{\partial x} = -\frac{\partial u}{\partial y} = \frac{2xy}{(x^2 + y^2)^2}.$$

The derivative is

$$\frac{\partial u}{\partial x} + i\frac{\partial v}{\partial x} = \frac{(y^2 - x^2) + 2ixy}{(x^2 + y^2)^2},$$

and we can check this answer by calculating that

$$-\frac{1}{z^2} = -\frac{\bar{z}^2}{(z\bar{z})^2} = \frac{(y^2 - x^2) + 2ixy}{(x^2 + y^2)^2}.$$

$\square$

Example 4.8

Show that the function $z \mapsto |z|^2$ is differentiable only at 0.

Solution

Here $u(x, y) = x^2 + y^2$ and $v(x, y) = 0$. Since $\partial u/\partial x = 2x$ and $\partial u/\partial y = 2y$, the Cauchy–Riemann equations are satisfied only at $z = 0$. Hence differentiability fails at all non-zero points. To verify differentiability at 0, observe that

$$\frac{|z|^2 - |0|^2}{z - 0} = \frac{z\bar{z}}{z} = \bar{z} \to 0$$

as $z \to 0$. The derivative exists, and has value 0. $\square$

Given that in real analysis the functions $x \mapsto x^2$ and $x \mapsto |x|^2$ are identical, the very different conclusions of Examples 4.6 and 4.8 are, on the face of it, surprising. One way of explaining the crucial difference between them is to observe that z^2 depends only on z, whereas $|z|^2 = z\bar{z}$ depends on both z and $\bar{z}$. Exploring this idea, we express the complex function $f(x, y) = u(x, y) + iv(x, y)$ of two real variables x and y as a function of z and $\bar{z}$, using the equations

$$x = \frac{1}{2}(z + \bar{z}), \quad y = \frac{1}{2i}(z - \bar{z}).$$

If we then pretend that we can apply standard two-variable calculus to this situation, we find that

$$\frac{\partial f}{\partial \bar{z}} = \frac{\partial f}{\partial x}\frac{\partial x}{\partial \bar{z}} + \frac{\partial f}{\partial y}\frac{\partial y}{\partial \bar{z}} = \frac{1}{2}\left(\frac{\partial f}{\partial x} + i\frac{\partial f}{\partial y}\right),$$

and this is equal to 0 (so that f is a function of z only) if and only if $\partial f/\partial x + i\partial f/\partial y = 0$, that is, if and only if

$$\left(\frac{\partial u}{\partial x} - \frac{\partial v}{\partial y}\right) + i\left(\frac{\partial v}{\partial x} + \frac{\partial u}{\partial y}\right) = 0,$$

that is, if and only if the Cauchy–Riemann equations are satisfied. This far from rigorous argument helps to convince us that the Cauchy–Riemann equations mark the difference between a true function of z and something that is merely a complex-valued function of two real variables.

From the Mean Value Theorem of real analysis (see [9, Theorems 4.7 and 4.8]) we know that a function whose derivative is 0 throughout (a, b) has a constant value. A similar result holds for holomorphic complex functions.

Theorem 4.9

Let f be holomorphic in a neighbourhood $U = N(a_0, R)$, and suppose that $f'(z) = 0$ for all z in U. Then f is constant.

Proof

In the usual way, let $f(x + iy) = u(x, y) + iv(x, y)$. Then

$$f'(z) = \frac{\partial u}{\partial x} + i\frac{\partial v}{\partial x} = \frac{\partial v}{\partial y} - i\frac{\partial u}{\partial y},$$

and so $f'(z) = 0$ implies that

$$\frac{\partial u}{\partial x} = \frac{\partial v}{\partial x} = \frac{\partial u}{\partial y} = \frac{\partial v}{\partial y} = 0$$

at every point in U.

Let $p = a + bi$ and $q = c + di$ be points in U. Then

at least one of $r = a + di$ and $s = c + bi$ lies in U. (See Exercise 4.2 below.) Suppose, without essential loss of generality, that r lies inside U. Then both $x \mapsto u(x, d)$ and $y \mapsto u(a, y)$ are real functions with zero derivative, and so are constant. Thus

$$u(a, b) = u(a, d) = u(c, d),$$

and similarly $v(a, b) = v(c, d)$. Thus f is constant.   □

Remark 4.10

This theorem can certainly be extended to a subset more general than an open disc, but not to a general open set. For example, if the open set U is a union of two disjoint open discs D_1 and D_2,

then the function

$$f(z) = \begin{cases} 1 & \text{if } x \in D_1 \\ 2 & \text{if } z \in D_2 \end{cases}$$

has zero derivative throughout its domain. This is an example where a more precise general theorem would require some exploration of topological ideas (the illustrated set U is not "connected"). In a first course, however, it is not a good idea to introduce too many new concepts, and I intend that all the sets we consider should have the connectedness property.

The proof of Theorem 4.9 involves a slightly clumsy double use of the real Mean Value Theorem, and it is natural to ask whether there is a useful complex analogue. One's first guess at such an analogue might be something like this: given a holomorphic function f and a pair of points c and w in its domain, then $f(w) = f(c) + (w - c)f'(\zeta)$, where ζ lies on the line segment between c and w. But this cannot be true in general. Suppose, for example, that $c = 1$, $w = i$ and $f(z) = z^3$. Then the proposed theorem would require the existence of ζ on the line segment between 1 and i with the property that $-i = 1 + 3(i - 1)\zeta^2$. Thus

$$\zeta^2 = \frac{-i - 1}{3(i - 1)} = \frac{i}{3}$$

and so $\zeta = \pm(1/\sqrt{3})e^{i\pi/4}$. For neither of these values does ζ lie on the line segment between 1 and i.

What we do have is a similar but less precise result, sometimes known as Goursat's[1] Lemma. Its proof is trivial, but we shall see that it is nonetheless a

---

[1] Edouard Jean-Baptiste Goursat, 1858–1936.

useful observation.

Theorem 4.11 (Goursat's Lemma)

Let f be holomorphic in an open subset U of $\mathbb{C}$, and let $c \in U$. Then there exists a function v with the property that

$$f(z) = f(c) + (z - c)f'(c) + (z - c)v(z, c), \qquad (4.4)$$

where $v(z, c) \to 0$ as $z \to c$.

Proof

Let

$$v(z, c) = \frac{f(z) - f(c)}{z - c} - f'(c).$$

Since f is holomorphic, $v(z, c) \to 0$ as $z \to c$, and Equation (4.4) clearly follows. □

The following alternative formulation of the lemma is perhaps worth recording here: if f is holomorphic in an open subset containing c, then there exists v such that

$$f(c + h) - f(c) - hf'(c) = hv(c, h), \qquad (4.5)$$

and $v(c, h) \to 0$ as $h \to 0$.

There is a converse to Goursat's Lemma:

Theorem 4.12

Let f be a function defined in an open subset U of $\mathbb{C}$, and let $c \in U$. If there exists a complex number A such that

$$\frac{f(z) - f(c) - A(z - c)}{z - c} \to 0 \quad \text{as} \ z \to c, \qquad (4.6)$$

then f is differentiable at c, and $f'(c) = A$.

Proof

From (4.6), with $z \neq c$, we deduce that

$$\frac{f(z) - f(c)}{z - c} - A \to 0$$

as $z \to c$, and the result follows immediately. □

We end this section with a striking result that is an easy consequence of Theorem 4.9:

Theorem 4.13

Let f be holomorphic in $N(a_0, R)$. If $|f|$ is constant in $N(a_0, R)$, then so is f.

Proof

If $|f| = 0$, then certainly $f = 0$. Suppose that $|f(z)| = c > 0$ for all z in $N(a_0, R)$. Thus

$$f(x + iy) = u(x, y) + iv(x, y),$$

and

$$[u(x, y)]^2 + [v(x, y)]^2 = c^2$$

for all $x + iy$ in $N(a_0, R)$. Hence, differentiating with respect to x and y, we have

$$uu_x + vv_x = 0, \quad uu_y + vv_y = 0.$$

From these equalities and the Cauchy–Riemann equations we deduce that

$$u^2 u_x = -uvv_x = uvu_y = -v^2 v_y = -v^2 u_x$$

$$u^2 v_x = -u^2 u_y = uvv_y = uvu_x = -v^2 v_x,$$

and from these equalities we deduce that

$$u^2 u_y = -u^2 v_x = v^2 v_x = -v^2 u_y,$$

$$u^2 v_y = u^2 u_x = -v^2 u_x = -v^2 v_y.$$

Thus

$$c^2 u_x = c^2 u_y = c^2 v_x = c^2 v_y = 0,$$

and so

$$u_x = u_y = v_x = v_y = 0.$$

Hence, by Theorem 4.9, f is constant throughout U.   □

EXERCISES

4.1. Verify the Cauchy–Riemann equations for the functions

    a) $f(z) = iz^2 + 2z$;

    b) $f(z) = (z + i)/(2z - 3i)$.

4.2. Let $p = a + ib$, $q = c + id$ lie within the neighbourhood $N(0, R)$. Show that at least one of $a + id$, $c + ib$ lies within $N(0, R)$.

4.3 Show that f is differentiable at c if and only if there is a function A, continuous at c, such that $f(z) = f(c) + A(z)(z - c)$. If this holds, show that $\lim_{z \to c} A(z) = f'(c)$.

4.2 Power Series

Infinite series involving real numbers play an important part in any first course in analysis, and some familiarity with the main ideas is assumed. (See Chapter 1.) Many of the essential definitions and results are easily extended to series involving complex numbers. For example, if $\sum_{n=1}^{N} z_n = S_N$, and if $S_N \to S$ as $N \to \infty$, then we say that the infinite series $\sum_{n=1}^{\infty} z_n$ **converges** to S, or that it has **sum** S. Again, it is just as true in the complex field as in the real field that if $\sum_{n=1}^{\infty} z_n$ is convergent then $\lim_{n \to \infty} z_n = 0$.

If $\sum_{n=1}^{\infty} |z_n|$ is convergent we say that $\sum_{n=1}^{\infty} z_n$ is **absolutely convergent**, and we may show, exactly as in the real case (see [9, Theorem 2.36]), that $\sum_{n=1}^{\infty} z_n$ is itself convergent. Since $\sum_{n=1}^{\infty} |z_n|$ is a series of (real) positive terms, the standard tests for convergence, such as the Comparison Test and the Ratio Test, can be applied. (See [9].)

Power series $\sum_{n=0}^{\infty} c_n(z - a)^n$, where z, a and c_n are complex, play a central role in complex analysis, and it is entirely appropriate, indeed necessary, to introduce them at this relatively early stage. The following simple theorem has some quite striking consequences:

Theorem 4.14

Suppose that the power series $\sum_{n=0}^{\infty} c_n(z - a)^n$ converges for $z - a = d$. Then it converges absolutely for all z such that $|z - a| < |d|$.

Proof

The series converges at all interior points of the circle.

Since $c_n d^n \to 0$ as $n \to \infty$, there exists $K > 0$ such that $|c_n d^n| \le K$ for all n. Let z be such that $|z - a| < |d|$. Then the geometric series $\sum_{n=0}^{\infty} \left(|z - a|/|d| \right)^n$ converges. Since, for all n,

$$|c_n(z - a)^n| = |c_n d^n| \left| \frac{z - a}{d} \right|^n \le K \left(\frac{|z - a|}{|d|} \right)^n ,$$

the series $\sum_{n=0}^{\infty} c_n(z - a)^n$ is, by the Comparison Test, (absolutely) convergent.

$\square$

As a consequence, we have an almost exact analogue of a result (Theorem 1.9) in real analysis:

Theorem 4.15

A power series $\sum_{n=0}^{\infty} c_n(z - a)^n$ satisfies exactly one of the following three conditions:

(i) the series converges for all z;

(ii) the series converges only for $z = a$;

(iii) there exists a positive real number R such that the series converges for all z such that $|z - a| < R$ and diverges for all z such that $|z - a| > R$.

Proof

Let $\mathcal{D}$ be the set of all z for which the series $\sum_{n=0}^{\infty} c_n(z - a)^n$ is convergent, and let

$$\mathcal{M} = \{ |z - a| : z \in \mathcal{D} \} .$$

Suppose first that $\mathcal{M}$ is unbounded. Then, for every z in $\mathbb{C}$, there exists d in $\mathcal{D}$ such that $|d| > |z - a|$, and it follows from Theorem 4.14 that $\sum_{n=0}^{\infty} c_n(z - a)^n$ is convergent. This is Case (i).

Suppose now that $\mathcal{M}$ is bounded, and let $R = \sup \mathcal{M}$. If $R = 0$ then $\mathcal{D} = \{a\}$, and we have Case (ii). So suppose that $R > 0$, and let z be such that $|z - a| < R$. Then, by definition of R, there exists d such that $|z - a| < |d| < R$, and such that $\sum_{n=0}^{\infty} c_n d^n$ is convergent. From Theorem 4.14 it follows that $\sum_{n=0}^{\infty} c_n (z - a)^n$ is convergent.

Now let z be such that $|z - a| > R$, and suppose that $\sum_{n=0}^{\infty} c_n (z - a)^n$ is convergent. Then $z - a \in \mathcal{D}$, and so we have an immediate contradiction, since R was to be an upper bound of $\mathcal{M}$. $\square$

The number R is called the **radius of convergence** of the power series, and we absorb Cases (i) and (ii) into this definition by writing $R = \infty$ for Case (i) and $R = 0$ for Case (ii).

Theorem 4.15 is silent concerning numbers z for which $|z - a| = R$, and this is no accident, for it is not possible to make a general statement. The circle $|z - a| = R$ is called the **circle of convergence**, but in using this terminology we are not implying that convergence holds for all – or indeed any – of the points on the circle.

The strong similarity between real and complex power series continues, for Theorem 1.9 extends to the complex case, and the proof is not significantly different. (See [9, Theorems 7.26 and 7.28].)

Theorem 4.16

Let $\sum_{n=0}^{\infty} c_n (z - a)^n$ be a power series with radius of convergence R.

(i) If
$$\lim_{n \to \infty} \left| \frac{c_n}{c_{n+1}} \right| = \lambda,$$
then $\lambda = R$.

(ii) If
$$\lim_{n \to \infty} |c_n|^{-1/n} = \lambda,$$
then $\lambda = R$.

$\square$

It is often convenient to prove results about power series for the case where $a = 0$, since it simplifies the notation, and it is easy to modify the results to cope with the general case, simply by substituting $z - a$ for z. The following useful theorem is a case in point. It tells us that if we differentiate a power series term by term we do not change the radius of convergence.

Theorem 4.17

The power series $\sum_{n=0}^{\infty} c_n (z - a)^n$ and $\sum_{n=1}^{\infty} n c_n (z - a)^{n-1}$ have the same radius of convergence.

Proof*

We shall prove this for the case where $a = 0$.

Suppose that the series $\sum_{n=0}^{\infty} c_n z^n$ and $\sum_{n=1}^{\infty} n c_n z^{n-1}$ have radii of convergence R_1, R_2, respectively. For each $z \neq 0$ and for all $n \geq 1$,

$$|c_n z^n| \leq |z| |n c_n z^{n-1}|,$$

and so, by the comparison test, $\sum_{n=0}^{\infty} c_n z^n$ is absolutely convergent for every z with the property that $\sum_{n=1}^{\infty} n c_n z^{n-1}$ is absolutely convergent, that is, for every z such that $|z| < R_2$. It follows that $R_2 \leq R_1$.

Suppose now, for a contradiction, that $R_2 < R_1$, and let z_1, z_2 be such that

$$R_2 < |z_2| < |z_1| < R_1.$$

From Exercise 2.12 we have

$$n \left| \frac{z_2}{z_1} \right|^{n-1} < \frac{|z_1|}{|z_1| - |z_2|},$$

and from this we deduce that, for all $n \geq 2$,

$$|n c_n z_2^{n-1}| < \frac{1}{|z_1| - |z_2|} |c_n z_1^n|.$$

Since $|z_1| < R_1$, the series $\sum_{n=0}^{\infty} |c_n z_1^n|$ is convergent. Hence, by the comparison test, $\sum_{n=1}^{\infty} |n c_n z_2^{n-1}|$ converges also, and this is a contradiction, since $|z_2| > R_2$. We deduce that $R_1 = R_2$.   □

Remark 4.18

The theorem holds good for a series with zero or (more importantly) infinite radius of convergence.

The crucial importance of power series can be seen from the next result. It is quite awkward to prove, but easy enough to understand. It tells us that, within the circle of convergence, it is legitimate to differentiate a series term by term.

Theorem 4.19

Let $\sum_{n=0}^{\infty} c_n (z - a)^n$ be a power series with radius of convergence $R \neq 0$, and let

$$f(z) = \sum_{n=0}^{\infty} c_n (z - a)^n \quad (|z| < R).$$

Then f is holomorphic within the open disc $N(a, R)$, and

$$f'(z) = \sum_{n=1}^{\infty} nc_n(z - a)^{n-1}.$$

Proof*

Again, it will be sufficient to prove this for the case where $a = 0$.

Let $g(z) = \sum_{n=1}^{\infty} nc_n z^{n-1}$. From Theorem 4.17 we know that g is defined for $|z| < R$. We shall show that, within the disc $N(a, R)$,

$$\left| \frac{f(z + h) - f(z)}{h} - g(z) \right| \to 0 \text{ as } h \to 0,$$

from which we deduce that f is differentiable, with derivative g.

The proof, though conceptually simple, is technically awkward, and it pays to record some preliminary observations. Let $0 < \rho < R$, and let z, h be such that $|z| < \rho$, $|z| + |h| < \rho$. Then the geometric series

$$\sum_{n=0}^{\infty} \frac{|z|^n}{\rho^n} \quad \text{and} \quad \sum_{n=0}^{\infty} \frac{(|z| + |h|)^n}{\rho^n}$$

are both convergent, with sums

$$\frac{\rho}{\rho - |z|}, \quad \frac{\rho}{\rho - |z| - |h|}, \tag{4.7}$$

respectively. Also, from Exercise 2.11 we know that

$$\sum_{n=0}^{\infty} nz^{n-1} = \frac{1}{(1 - z)^2} \quad (|z| < 1),$$

and it follows that

$$\sum_{n=0}^{\infty} \frac{n|z|^{n-1}}{\rho^n} = \frac{\rho}{(\rho - |z|)^2}. \tag{4.8}$$

We have that

$$\left| \frac{f(z + h) - f(z)}{h} - g(z) \right| = \left| \sum_{n=0}^{\infty} c_n \left(\frac{(z + h)^n - z^n}{h} - nz^{n-1} \right) \right|$$

$$\leq \sum_{n=0}^{\infty} |c_n| \left| \frac{(z + h)^n - z^n}{h} - nz^{n-1} \right|.$$

Now, by the binomial theorem,

$$\left| \frac{(z+h)^n - z^n}{h} - nz^{n-1} \right| = \left| \binom{n}{2} z^{n-2} h + \binom{n}{3} z^{n-3} h^2 + \cdots + h^{n-1} \right|$$

$$\leq \binom{n}{2} |z|^{n-2} |h| + \binom{n}{3} |z|^{n-3} |h|^2 + \cdots + |h|^{n-1}$$

$$= \frac{\left(|z| + |h| \right)^n - |z|^n}{|h|} - n|z|^{n-1} .$$

Also, since $c_n \rho^n \to 0$ as $n \to \infty$, there exists $K > 0$ such that, for all $n \geq 1$,

$$|c_n \rho^n| \leq K .$$

Hence

$$|c_n z^n| = |c_n \rho^n| \frac{|z|^n}{\rho^n} \leq K \frac{|z|^n}{\rho^n} ,$$

$$|c_n (z+h)^n| = |c_n \rho^n| \frac{|z+h|^n}{\rho^n} \leq K \frac{|z+h|^n}{\rho^n} \leq K \frac{\left(|z| + |h| \right)^n}{\rho^n} .$$

Hence,

$$\left| \frac{f(z+h) - f(z)}{h} - g(z) \right| \leq \sum_{n=0}^{\infty} \frac{K}{|h|} \left(\frac{\left(|z| + |h| \right)^n}{\rho^n} - \frac{|z|^n}{\rho^n} - \frac{n|h||z|^{n-1}}{\rho^n} \right)$$

$$= \frac{K}{|h|} \left(\frac{\rho}{\rho - |z| - |h|} - \frac{\rho}{\rho - |z|} - \frac{\rho|h|}{\left(\rho - |z| \right)^2} \right) \quad \text{(by (4.7) and (4.8))}$$

$$= \frac{K\rho}{|h|} \left(\frac{|h|}{\left(\rho - |z| - |h| \right) \left(\rho - |z| \right)} - \frac{|h|}{\left(\rho - |z| \right)^2} \right)$$

$$= \frac{K\rho}{\rho - |z|} \left(\frac{1}{\left(\rho - |z| - |h| \right)} - \frac{1}{\rho - |z|} \right)$$

$$= \frac{K\rho|h|}{\left(\rho - |z| - |h| \right) \left(\rho - |z| \right)^2} ,$$

and this tends to 0 as $h \to 0$.   □

Example 4.20

Sum the series

$$1^2 + 2^2 z + 3^2 z^2 + 4^2 z^3 + \cdots \quad (|z| < 1) .$$

Solution

From
$$1 + z + z^2 + z^3 + \cdots = \frac{1}{1-z} \quad (|z| < 1)$$

we deduce, by differentiating term by term, that
$$1 + 2z + 3z^2 + 4z^3 + \cdots = \frac{1}{(1-z)^2} \quad (|z| < 1).$$

Hence
$$z + 2z^2 + 3z^3 + 4z^4 + \cdots = \frac{z}{(1-z)^2} \quad (|z| < 1),$$

and so, again by differentiation, for all z in $N(0,1)$,
$$1^2 + 2^2 z + 3^2 z^2 + 4^2 z^3 + \cdots = \frac{d}{dz}\left(\frac{z}{(1-z)^2}\right) = \frac{1+z}{(1-z)^3}.$$

We define the function exp by means of a power series, convergent for all z:
$$\exp z = \sum_{n=0}^{\infty} \frac{z^n}{n!} = 1 + z + \frac{z^2}{2!} + \frac{z^3}{3!} + \cdots. \tag{4.9}$$

The function is holomorphic over the whole complex plane, and one easily verifies that
$$(\exp)'(z) = \exp z. \tag{4.10}$$

Let $F_w(z) = \exp(z + w)/\exp z$. Then, by the quotient rule,
$$F_w'(z) = \frac{(\exp z)\big(\exp(z+w)\big) - \big(\exp(z+w)\big)(\exp z)}{(\exp z)^2} = 0,$$

and so $F_w(z) = k$, a constant. (See Exercise 3.2.) Since $F_w(0) = \exp w$, we deduce that $F_w(z) = \exp w$ for all z, and so we have the crucial property of the exponential function, that
$$\exp(z + w) = (\exp z)(\exp w). \tag{4.11}$$

In real analysis (see [9, Chapter 6]) we use this property to establish, for every rational number q, that $\exp q = e^q$, where $e = \exp 1$, and then we **define** e^x to be $\exp x$ for every real number x. It is equally reasonable to **define** e^z to be $\exp z$ for all z in $\mathbb{C}$. We shall use both notations. The functions cos, sin, cosh and sinh defined by

$$\cos z = \sum_{n=0}^{\infty}(-1)^n \frac{z^{2n}}{(2n)!} = 1 - \frac{z^2}{2!} + \frac{z^4}{4!} - \cdots , \tag{4.12}$$

$$\sin z = \sum_{n=0}^{\infty}(-1)^n \frac{z^{2n+1}}{(2n+1)!} = z - \frac{z^3}{3!} + \frac{z^5}{5!} - \cdots , \tag{4.13}$$

$$\cosh z = \sum_{n=0}^{\infty} \frac{z^{2n}}{(2n)!} = 1 + \frac{z^2}{2!} + \frac{z^4}{4!} + \cdots , \tag{4.14}$$

$$\sinh z = \sum_{n=0}^{\infty} \frac{z^{2n+1}}{(2n+1)!} = z + \frac{z^3}{3!} + \frac{z^5}{5!} + \cdots , \tag{4.15}$$

are all entire functions (holomorphic over the whole complex plane), and it is easy to verify that the formulae

$$\cos z + i\sin z = e^{iz}, \qquad \cosh z + \sinh z = e^z , \tag{4.16}$$

$$\cos z = \frac{1}{2}(e^{iz} + e^{-iz}), \qquad \sin z = \frac{1}{2i}(e^{iz} - e^{-iz}), \tag{4.17}$$

$$\cos^2 z + \sin^2 z = 1 , \tag{4.18}$$

$$\cosh z = \frac{1}{2}(e^z + e^{-z}), \qquad \sinh z = \frac{1}{2}(e^z - e^{-z}), \tag{4.19}$$

$$(\cos)'(z) = -\sin z , \qquad (\sin)'(z) = \cos z , \tag{4.20}$$

$$(\cosh)'(z) = \sinh z , \qquad (\sinh)'(z) = \cosh z , \tag{4.21}$$

are valid for all complex numbers z.

It is not by any means obvious that for all real x the sine and cosine defined by means of these power series are the same as the geometrically defined sine and cosine that enable us to put complex numbers into polar form. A proof that they are in fact the same can be found in [9, Chapter 8]. Here we shall **assume** that the functions cos and sin, defined by the above power series, have the properties

$$\cos x > 0 \; (x \in [0, \pi/2)), \quad \cos(\pi/2) = 0 .$$

(In a strictly logical development of analysis, this is the *definition* of π. See [9, Chapter 8].) From (4.18) we deduce that

$$\sin(\pi/2) = \pm\sqrt{1 - \cos^2(\pi/2)} = \pm 1 .$$

Since $\sin 0 = 0$ and $(\sin)'(x) = \cos x > 0$ in $[0, \pi/2)$, we must in fact have $\sin(\pi/2) = 1$. From (4.11) and (4.16) we see that

$$\cos(z + w) + i\sin(z + w) = e^{i(z+w)} = e^{iz}e^{iw}$$
$$= (\cos z + i\sin z)(\cos w + i\sin w)$$
$$= (\cos z \cos w - \sin z \sin w) + i(\sin z \cos w + \cos z \sin w) ,$$

and so the familiar addition formulae

$$\cos(z + w) = \cos z \cos w - \sin z \sin w \qquad (4.22)$$
$$\sin(z + w) = \sin z \cos w + \cos z \sin w \qquad (4.23)$$

hold for all z, w in $\mathbb{C}$. From these it follows that

$$\cos \pi = \cos^2(\pi/2) - \sin^2(\pi/2) = -1, \quad \sin \pi = 2 \sin(\pi/2) \cos(\pi/2) = 0,$$

$$\cos 2\pi = \cos^2 \pi - \sin^2 \pi = 1, \quad \sin 2\pi = 2 \sin \pi \cos \pi = 0.$$

Hence we have the periodic property of the exponential function: for all z in $\mathbb{C}$,

$$e^{z+2\pi i} = e^z(\cos 2\pi + i \sin 2\pi) = e^z. \qquad (4.24)$$

Writing z as $x + iy$ with x, y in $\mathbb{R}$, we see that

$$e^z = e^{x+iy} = e^x e^{iy} = e^x(\cos y + i \sin y).$$

Thus

$$|e^z| = e^x, \quad \arg e^z \equiv y \pmod{2\pi}. \qquad (4.25)$$

Since e^x is non-zero for all real x, we have the important conclusion that e^z is non-zero for all complex numbers z. Thus $z \mapsto e^{-z} = 1/e^z$ is also an entire function.

EXERCISES

4.4. Let $p(z) = a_0 + a_1 z + a_2 z^2 + \cdots + a_n z^n$ be a polynomial of degree n. Show that
$$\lim_{|z| \to \infty} \frac{p(z)}{a_n z^n} = 1.$$

4.5. Show that $\overline{e^z} = e^{\bar{z}}$ for all z in $\mathbb{C}$, and deduce that
$$\overline{\sin z} = \sin \bar{z}, \quad \overline{\cos z} = \cos \bar{z}.$$

4.6. Use Formulae (4.17) to prove that, for all complex numbers z, w
$$\cosh(z + w) = \cosh z \cosh w + \sinh z \sinh w,$$
$$\sinh(z + w) = \sinh z \cosh w + \cosh z \sinh w.$$

4.7. Show that, if $F(z) = \cosh^2 z - \sinh^2 z$, then $F'(z) = 0$, and deduce that $\cosh^2 z - \sinh^2 z = 1$ for all z.

4.8. Show that $\cos(iz) = \cosh z$ and $\sin(iz) = i \sinh z$. Determine the real and imaginary parts of $\cos z$ and $\sin z$, and verify the Cauchy–Riemann equations for each of the functions cos and sin.

Show that

$$| \sin z|^2 = \sin^2 x + \sinh^2 y, \quad | \cos z|^2 = \cos^2 x + \sinh^2 y.$$

4.9. We know that, for all z in $\mathbb{C}$,

$$\cos^2 z + \sin^2 z = 1.$$

This does **not** imply that $| \cos z| \leq 1$ and $| \sin z| \leq 1$. Show in fact that, for all real y,

$$| \cos(iy)| > \frac{1}{2} e^{|y|}, \quad | \sin(iy)| \geq \frac{1}{2} (e^{|y|} - 1).$$

4.10. Show that, for all z in $\mathbb{C}$ and all n in $\mathbb{Z}$,

$$\sin(z + n\pi) = (-1)^n \sin z, \qquad \cos(z + n\pi) = (-1)^n \cos z.$$

4.11. Show that, for all z in $\mathbb{C}$,

$$\cosh(z + 2\pi i) = \cosh z \qquad \sinh(z + 2\pi i) = \sinh z.$$

4.12. Determine the real and imaginary parts of $\exp(z^2)$ and $\exp(\exp z)$.

4.13. Show that, if x and y are real,

$$| \sin(x + iy)| \geq \sinh y.$$

4.14. Show that, as $z \to 0$,

$$e^z = 1 + z + O(z^2), \quad \cos z = 1 - \frac{z^2}{2} + o(z^3).$$

4.3 Logarithms

In real analysis the statements $y = e^x$ and $x = \log y$ (where $y > 0$) are equivalent. (Here log is of course the natural logarithm, to the base e.) If we try to use this approach to define $\log z$ for complex z then we hit a difficulty, for the fact that $e^z = e^{z+2\pi i}$ for all z means that $z \mapsto e^z$ is no longer a one-to-one function. The notion of a logarithm is indeed useful in complex analysis, but we have to be careful. Let us suppose that $w = \log z$ (where $z \neq 0$) is equivalent to $z = e^w$, where $w = u + iv$. Then

$$z = e^{u+iv} = e^u e^{iv},$$

and so $e^u = |z|$, $v \equiv \arg z \pmod{2\pi}$. Thus $u = \log|z|$, while v is defined only modulo 2π. The **principal logarithm** is given by

$$\log z = \log|z| + i \arg z,$$

where $\arg z$ is the principal argument, lying in the interval $(-\pi, \pi]$. It is convenient also to refer to the value of the principal logarithm at z as the **principal value** of the logarithm at z. It should be emphasised that the choice of the principal argument as lying in $(-\pi, \pi]$ is completely arbitrary: we might have chosen the interval $[0, 2\pi)$ – or indeed the interval $(-\pi/8, 15\pi/8]$ – instead. It follows that the choice of the principal logarithm is similarly arbitrary. However, we have made a choice, and we shall stick to it.

With the choice we have made,

$$\log(-1) = i\pi, \quad \log(-i) = -i(\pi/2), \quad \log(1 + i\sqrt{3}) = \log 2 + i(\pi/3),$$

and so on. Statements such as

$$\log(z_1 z_2) = \log z_1 + \log z_2 \quad \text{and} \quad \log(z^k) = k \log z$$

need to be treated with some care, for the imaginary parts may differ by a multiple of 2π. To take the simplest example, $\log(-1) + \log(-1) = 2i\pi$, and this is *not* the principal logarithm of $(-1)(-1)$.

A useful approach to the untidiness caused by functions such as arg and log is to define a **multifunction** f as a rule associating each z in its domain with a *subset* of $\mathbb{C}$. The elements of the subset are called the **values** of the multifunction. Thus we can define $\operatorname{Arg} z$ (note the capital letter) as $\{\arg z + 2n\pi : n \in \mathbb{Z}\}$, and $\operatorname{Log} z$ as $\{\log z + 2n\pi i : n \in \mathbb{Z}\}$. Then we can say definitely that

$$\operatorname{Arg}(zw) = \operatorname{Arg} z + \operatorname{Arg} w, \quad \operatorname{Log}(zw) = \operatorname{Log} z + \operatorname{Log} w,$$

where, for example, the second statement means that every value of the multifunction $\text{Log}(zw)$ is a sum of a value of $\text{Log}\,z$ and a value of $\text{Log}\,w$; and, conversely, the sum of an arbitrary value of $\text{Log}\,z$ and an arbitrary value of $\text{Log}\,w$ is a value of $\text{Log}(zw)$.

The multifunctional nature of the logarithm affects the meaning of powers c^z, where $c, z \in \mathbb{C}$. We define c^z in the obvious way as $e^{z \log c}$, and immediately realise that $z \mapsto c^z$ may sometimes have to be interpreted as a multifunction. If we use the principal logarithm of c we can assure ourselves that $c^z c^w = c^{z+w}$, but $(c^z)^w = c^{zw}$ and $c^z d^z = (cd)^z$ may fail unless we interpret them in multifunction mode.

Example 4.21

Describe $(1 + i)^i$.

Solution

This is not a single complex number, but a set:

$$(1 + i)^i = e^{i\,\text{Log}(1+i)} = \left\{ \exp\left[i\left(\log(\sqrt{2}) + (2n + \tfrac{1}{4})\pi i \right) \right] \; : \; n \in \mathbb{Z} \right\}$$
$$= \left\{ \exp\left[-(2n + \tfrac{1}{4})\pi + i \log(\sqrt{2}) \right] \; : \; n \in \mathbb{Z} \right\}.$$

$\square$

Example 4.22

Comment on the statements

$$e^{\log z} = z, \quad \log(e^z) = z, \quad (e^z)^w = e^{zw}.$$

Solution

For the first formula, given the ambiguity of log, we should examine the multifunction $e^{\text{Log}\,z}$. However, we find that

$$e^{\text{Log}\,z} = \left\{ e^{\log|z| + i \arg z + 2n\pi i} \; : \; n \in \mathbb{Z} \right\} = \left\{ |z| e^{i \arg z} e^{2n\pi i} \; : \; n \in \mathbb{Z} \right\}$$
$$= \left\{ z e^{2n\pi i} \; : \; n \in \mathbb{Z} \right\} = \{z\},$$

and so the first formula can be used with perfect safety.

On the same principle, we next examine the multifunction $\text{Log}(e^z)$. Here $w \in \text{Log}(e^z)$ if and only if $e^w = e^z$, that is, if and only if

$$w \in \{ z + 2n\pi i \; : \; n \in \mathbb{Z} \} = \{ x + (y + 2n\pi)i \; : \; n \in \mathbb{Z} \}.$$

This set certainly **includes** z, but we cannot be sure that using the principal logarithm will give us the answer z. For example, if $z = 5i\pi/2$, then $\log(e^z) = i\pi/2 \neq z$.

By definition,

$$(e^z)^w = e^{w \operatorname{Log}(e^z)} = \{e^{w(z+2n\pi i)} : n \in \mathbb{Z}\} = \{e^{zw} e^{2n\pi iw} : n \in \mathbb{Z}\}.$$

In fact all we can say is that e^{zw} is a value of the multifunction $(e^z)^w$. $\qquad\square$

All this may seem somewhat confusing, but in practice it is usually surprisingly easy to sort out whether or not a formula is true in function or in multifunction mode. If a is real and positive, we shall normally regard $z \mapsto a^z$ as a function rather than a multifunction. Thus a^z is defined as $e^{z \log a}$, where log has its usual real analysis meaning.

Finally, we would expect that the formula for the differentiation of the real function $\log x$ might extend to the complex plane. Also, since all the values of the multifunction $\operatorname{Log} z$ differ by a constant, we would expect the ambiguity to disappear on differentiation:

Theorem 4.23

$$(\log)'(z) = \frac{1}{z}.$$

Proof

Let $z = x + iy = re^{i\theta}$. Then the values of $\operatorname{Log} z$ are given by

$$\{\log r + i(\theta + 2n\pi) : n \in \mathbb{Z}\}.$$

If we choose any one of these values, we see that

$$\log z = \frac{1}{2} \log(x^2 + y^2) + i(\tan^{-1}(y/x) + 2n\pi + C),$$

where $C = 0$ or $\pm\pi$. (The $\pm\pi$ is necessary, since $\tan^{-1}(y/x)$ by definition lies between $-\pi/2$ and $\pi/2$, and so, for example, $\arg(-1 - i) = -3\pi/4 = \tan^{-1} 1 - \pi$.) Hence, calculating the partial derivatives with respect to x of the real and imaginary parts, we see that

$$(\log)'(z) = \frac{x}{x^2 + y^2} + i\frac{1}{1 + (y/x)^2} \cdot \frac{-y}{x^2} = \frac{x - iy}{x^2 + y^2} = \frac{1}{z}.$$

$\qquad\square$

EXERCISES

4.15. Describe the multifunction z^i, and determine the real and imaginary parts of the multifunction $(-i)^i$.

4.16. Define the multifunction Sin^{-1} by the rule that $w \in \operatorname{Sin}^{-1} z$ if and only if $\sin w = z$. Show that

$$\operatorname{Sin}^{-1} z = -i \operatorname{Log}\left(iz \pm \sqrt{1 - z^2}\right).$$

Describe $\operatorname{Sin}^{-1}(1/\sqrt{2})$.

4.17. By analogy, define Tan^{-1} by the rule that $w \in \operatorname{Tan}^{-1} z$ if and only if $\tan w = z$. Show that

$$\operatorname{Tan}^{-1} z = \frac{1}{2i} \operatorname{Log}\left(\frac{1 + iz}{1 - iz}\right) \quad (z \neq \pm i).$$

Suppose now that $z = e^{i\theta}$, of modulus 1, where $-\pi/2 < \theta < \pi/2$. Show that

$$\operatorname{Tan}^{-1}(e^{i\theta}) = \frac{1}{2i}\left(\log\left|\frac{\cos\theta}{1 + \sin\theta}\right| + i\left(2n\pi + \frac{\pi}{2}\right)\right),$$

and deduce that

$$\operatorname{Re}\left(\operatorname{Tan}^{-1}(e^{i\theta})\right) = \left\{n\pi + \frac{\pi}{4} : n \in \mathbb{Z}\right\}.$$

4.18. Comment, on the mathematical rather than the literary content, of:

> Little Jack Horner sat in a corner
> Trying to work out π.
> He said, "It's the principal logarithm
> Of $(-1)^{-i}$."

4.4 Cuts and Branch Points

As we have seen in the exercises above, there are many multifunctions, and it is easy to define still more complicated examples. For our purposes only two multifunctions will really matter, namely $\operatorname{Log} z$ (along with its close companion $\operatorname{Arg} z$) and $z^{1/n}$, and there is an easy way of dealing with them. First, the principal logarithm is holomorphic in any region contained in $\mathbb{C} \setminus (-\infty, 0]$. We think of the plane as having a **cut** along the negative x-axis, preventing $\arg z$ from leaving the interval $(-\pi, \pi]$.

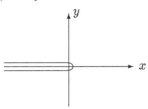

If z moves round any closed path wholly contained in $\mathbb{C}\backslash(-\infty, 0]$, the logarithm changes continuously and returns to its original value.

As with the definition of the principal argument and the principal logarithm, the position of the cut is ultimately arbitrary. The positive x-axis would do as well. So would any half-line containing the origin, and more complicated cuts would also be possible. The key points are that the cut should contain 0 and should go off to infinity, and we say that 0 and ∞ are **branch points**. If our cut failed to contain 0 or failed to go off to infinity without gaps we could find a circular path round which the logarithm could not both change continuously and return to its original value.

The other inescapable multifunction is $z^{1/n}$, where n is a positive integer. If $z = e^{i\theta}$ then, as a multifunction,

$$z^{1/n} = \{r^{1/n}e^{i(\theta + 2k\pi)/n} : k = 0, 1, \ldots, n-1\}.$$

Again the position of the cut is to an extent arbitrary, but the natural way to proceed is to define $r^{1/n}e^{i\theta/n}$ as the principal value and to make a cut along $[0, \infty)$. Once again, if z moves round any closed path wholly contained in $\mathbb{C} \backslash [0, \infty)$ then the value of $z^{1/n}$ (whether the principal value or not) changes continuously and returns to its original value. Again, 0 and ∞ are branch points.

For more complicated multifunctions it can be harder to determine the branch points and the appropriate cut, but the functions we have mentioned will be sufficient for our requirements.

4.5 Singularities

Let f be a complex function whose domain includes the neighbourhood $N(c, r)$. It can happen that $\lim_{z \to c} f(z)$ exists, but is not equal to $f(c)$. In such a case we say that f has a **removable singularity** at the point c. The terminology is apposite, for we can remove the singularity by redefining $f(c)$ as $\lim_{z \to c} f(z)$. For example, we might, admittedly somewhat perversely, define

$$f(z) = \begin{cases} z^2 & \text{if } z \neq 2 \\ 5 & \text{if } z = 2. \end{cases}$$

Then f has a removable singularity at 2, and the singularity disappears if we redefine $f(2)$ to be 4. Singularities of this kind play no significant part in the development of our theory, and when in future we refer to singularities, it will be assumed that they are not of this artificial kind.

More importantly, we have already come across a function, namely $z \mapsto 1/z$, which is holomorphic in any region not containing 0. For a complex function f, a point c such that $f(z)$ has no finite limit as $z \to c$ is called a **singularity**. If there exists $n \geq 1$ such that $(z-c)^n f(z)$ has a finite limit as $z \to c$, we say that the singularity is a **pole**. The **order** of the pole is the least value of n for which $\lim_{z \to c}(z-c)^n f(z)$ is finite. Poles of order 1, 2 and 3 are called (respectively) **simple**, **double** and **triple**. If f is a function holomorphic on an open subset H of $\mathbb{C}$ except possibly for poles, we say that f is **meromorphic in H**. It is, for example, clear that the function $1/z$ is meromorphic (in $\mathbb{C}$), with a simple pole at 0.

Example 4.24

Show that $1/\sin z$ is meromorphic in $\mathbb{C}$, with simple poles at $z = n\pi$ $(n \in \mathbb{Z})$.

Solution

From Exercise 3.8 we know that

$$\sin(x + iy) = \sin x \cosh y + i \cos x \sinh y,$$

and we know that $\cosh y \geq 1$ for all real y, and $\sinh y = 0$ if and only if $y = 0$. Hence $\text{Re}(\sin(x + iy)) = 0$ if and only if $\sin x = 0$, that is, if and only if $x = n\pi$. Since $\cos n\pi = \pm 1$, $\text{Im}(\sin(x + iy)) = 0$ if and only if $y = 0$. Thus the singularities of the function $1/\sin z$ occur exactly at the points $n\pi$.

From Exercise 3.11 we know that $\sin z = (-1)^n \sin(z - n\pi)$ for all n in $\mathbb{Z}$. Hence

$$\lim_{z \to n\pi} \frac{z - n\pi}{\sin z} = \lim_{z \to n\pi} \frac{(-1)^n (z - n\pi)}{\sin(z - n\pi)} = (-1)^n.$$

$\square$

Example 4.25

Show that $\cos z = -1$ if and only if $z = (2n + 1)\pi$, where $n \in \mathbb{Z}$. Hence show that

$$\frac{1}{1 + \cosh z}$$

has double poles at $z = (2n + 1)\pi i$.

Solution

One way round the first statement is clear: we know that $\cos(2n + 1)\pi = -1$. For the converse, note that $\cos(x + iy) = -1$ if and only if

$$\cos x \cosh y - i \sin x \sinh y = -1 \,,$$

that is, if and only if

$$\cos x \cosh y = -1 \,, \tag{4.26}$$

$$\sin x \sinh y = 0 \,. \tag{4.27}$$

From (4.27) we deduce that either (i) $y = 0$ or (ii) $x = m\pi$ (where $m \in \mathbb{Z}$). Suppose first that $y = 0$. Then $\cosh y = 1$ and so, from (4.26), $\cos x = -1$. Hence $x + iy = (2n + 1)\pi + 0i$, as required. Next, suppose that $x = m\pi$. Then $\cos x = \pm 1$, and so (4.26) gives $(\pm 1) \cosh y = -1$. Since $\cosh y > 1$ for all $y \neq 0$, this can happen only if $y = 0$ and $\cos x = -1$, that is, only if $x + iy = (2n + 1)\pi + 0i$.

Turning now to the second part of the question, we begin by observing that $1 + \cosh z = 1 + \cos(iz)$, and so $1 + \cosh z = 0$ if and only if iz is an odd multiple of π, that is, if and only if $z = (2n + 1)\pi i$. So the singularities of $1/(1 + \cosh z)$ occur at these points. The periodicity of cos gives

$$\cosh z = \cos(iz) = -\cos\big(iz + (2n + 1)\pi\big)$$
$$= -\cos i\big(z - (2n + 1)\pi i\big) = -\cosh\big(z - (2n + 1)\pi i\big) \,,$$

and so, as $z \to (2n + 1)\pi i$,

$$\frac{\big(z - (2n + 1)\pi i\big)^k}{1 + \cosh z} = \frac{\big(z - (2n + 1)\pi i\big)^k}{1 - \cosh\big(z - (2n + 1)\pi i\big)}$$

$$= \big(z - (2n + 1)\pi i\big)^k \bigg/ \bigg[-\frac{\big(z - (2n + 1)\pi i\big)^2}{2!} - \frac{\big(z - (2n + 1)\pi i\big)^4}{4!} - \cdots \bigg]$$

$$\to \begin{cases} \infty & \text{if } k = 1 \\ -2 & \text{if } k = 2. \end{cases}$$

Thus the singularities are all double poles. $\qquad\square$

If p and q are polynomial functions, we say that the function r, defined on the domain $\{z \in \mathbb{C} : q(z) \neq 0\}$ by $r(z) = p(z)/q(z)$, is a **rational** function. If we suppose, without essential loss of generality, that p and q have no common factors, then r is a meromorphic function with poles at the roots of the equation $q(z) = 0$. For example, $z \mapsto (z+1)/z(z-1)^2$ has a simple pole at $z = 0$ and a double pole at $z = 1$.

Other types of singularity can arise, and will be discussed properly later. For example, the function $e^{1/z}$ clearly has a singularity at $z = 0$, but this is not a pole, since for all $n \geq 1$

$$z^n e^{1/z} = z^n \left(1 + \frac{1}{z} + \cdots + \frac{1}{(n+1)!z^{n+1}} + \cdots\right)$$

has no finite limit as $z \to 0$. This is an example of an **isolated essential singularity**. Even worse is $\tan(1/z)$, which has a sequence $(2/n\pi)_{n\in\mathbb{N}}$ of singularities (in fact poles) with limit 0. At 0 we have what is called a **non-isolated essential singularity**.

EXERCISES

4.19. Show that $z \mapsto \tan z$ is meromorphic, with simple poles at $(2n + 1)\pi/2$ $(n \in \mathbb{Z})$.

4.20. Investigate the singularities of $z \mapsto 1/(z \sin z)$.

4.21. Let r be a rational function with a pole of order k at the point c. Show that the derivative of r has a pole of order $k + 1$ at c.

5
Complex Integration

5.1 The Heine–Borel Theorem

The rather technical Heine[1]–Borel[2] Theorem is necessary for some of our proofs, and this is as good a place as any to introduce it. The result we shall need most immediately is Theorem 5.3.

A subset S in $\mathbb{C}$ is said to be **bounded** if there exists a positive constant K such that $|z| \leq K$ for all z in S. Geometrically speaking, S lies inside the closed disc $\bar{N}(0, K)$.

By an **open covering** of a set S we mean a possibly infinite collection

$$\mathcal{C} = \{V_i : i \in I\}$$

of open sets V_i whose union $\bigcup \{V_i : i \in I\}$ contains the set S. If I is finite we say that the covering is **finite**. A **subcovering** of $\mathcal{C}$ is a selection $\mathcal{S}$ of the open sets V_i which still has union containing S:

$$\mathcal{S} = \{V_i : i \in J\},$$

where J is a subset of I and $S \subseteq \bigcup \{V_i : i \in J\}$. It is called a **finite** subcovering if J is a finite subset of I.

---

[1] Heinrich Eduard Heine, 1821–1881.
[2] Félix Édouard Justin Émile Borel, 1871–1956.

Theorem 5.1 (The Heine–Borel Theorem)

Let S be a closed, bounded subset of $\mathbb{C}$. Then every open covering of S contains a finite subcovering of S.

Proof*

Since S is bounded, we may suppose that it is enclosed within a square Q with side l:

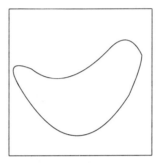

Let $\mathcal{C}$ be an open covering of S, and let us suppose, for a contradiction, that there is no finite subcollection of $\mathcal{C}$ that covers S. Divide the square Q into 4 equal parts by bisecting the sides. Then for at least one of these parts – call it Q_1 – the set $S \cap Q_1$ of S is not covered by a finite subcollection of $\mathcal{C}$. We may now subdivide the square Q_1 in the same way and obtain a still smaller square Q_2 with the property that the set $S \cap Q_2$ is not covered by a finite subcollection of $\mathcal{C}$. Continuing this argument we find squares

$$Q \supset Q_1 \supset Q_2 \supset Q_3 \supset \cdots,$$

with the property that $S \cap Q_n$ is not covered by any finite subcollection of $\mathcal{C}$. For $n = 1, 2, 3, \ldots$, Q_n is a square with side $l/2^n$, and the distance between any two points within Q_n is less than $l\sqrt{2}/2^n$, the length of the diagonal of the square.

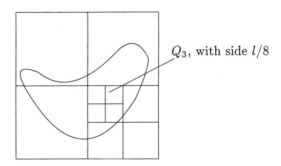

Q_3, with side $l/8$

For each n, let α_n be the centre of the square Q_n. We show that the sequence (α_n) is a Cauchy sequence. For a given $\epsilon > 0$ there exists N such that $l\sqrt{2}/2^N < \epsilon$. If $m, n > N$ then both α_m and α_n are inside or on the boundary of the square Q_N, and so

$$|\alpha_m - \alpha_n| \le l\sqrt{2}/2^N < \epsilon.$$

By the completeness property it follows that the sequence (α_n) has a limit α.

We show now that α lies inside or on the boundary of *every* square Q_N. Let $N \in \mathbb{N}$. Then, for all $m > N$ the point α_m lies inside or on the square Q_N, and so $|\alpha_m - \alpha_N| \le l\sqrt{2}/2^{N+1}$. It follows that

$$|\alpha - \alpha_N| = \lim_{m \to \infty} |\alpha_m - \alpha_N| \le l\sqrt{2}/2^{N+1},$$

and so α lies inside or on the boundary of Q_N.

Since $\mathcal{C}$ is a covering, $\alpha \in U$ for some U in $\mathcal{C}$. Since U is open, there exists a neighbourhood $N(\alpha, \delta)$ wholly contained in U. If we now choose n so that $l\sqrt{2}/2^n < \delta$, we see that the square Q_n lies entirely within $N(\alpha, \delta)$ and so entirely within U.

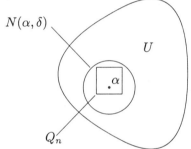

This is a contradiction, for Q_n, chosen so as to be covered by no finite selection of the open sets from $\mathcal{C}$, is in fact covered by the single open set U. $\qquad \square$

Remark 5.2

Both "closed" and "bounded" are required in the theorem. If, for example, S is the bounded open disc $N(0, 1)$, then

$$\mathcal{C} = \{N(0, 1 - \tfrac{1}{n}) : n \in \mathbb{N}\}$$

is an open covering of S, but no finite subcovering of $\mathcal{C}$ will suffice. Similarly, if S is the closed, unbounded set

$$\{z \in \mathbb{C} : \operatorname{Re} z \ge 0 , \operatorname{Im} z \ge 0\}$$

(the first quadrant of the complex plane), then

$$\mathcal{C} = \{N(0, n) : n \in \mathbb{N}\}$$

is an open covering of S, but again no finite subcovering will suffice.

One very significant consequence of the Heine–Borel Theorem is as follows:

Theorem 5.3

Let S be a closed, bounded set, and let f, with domain containing S, be continuous. Then f is bounded on S; that is, the set

$$\{|f(z)| : z \in S\}$$

is bounded. Moreover, if $M = \sup_S |f|$, then there exists z in S such that $|f(z)| = M$.

Proof*

Let $\epsilon = 1$. By continuity, for each c in S there exists $\delta_c > 0$ such that $|f(z) - f(c)| < 1$ for all z in $N(c, \delta_c)$. The sets $N(c, \delta_c)$ certainly cover S, and so, by the Heine–Borel Theorem, a finite subcollection

$$\{N(c_i, \delta_{c_i}) : i = 1, 2, \ldots, m\}$$

covers S. Let

$$K = \max \{|f(c_1)| + 1, \ldots, |f(c_m)| + 1\},$$

and let $z \in S$. Then $z \in N(c_i, \delta_{c_i})$ for at least one i in $\{1, 2, \ldots, m\}$, and so

$$|f(z)| - |f(c_i)| \le |f(z) - f(c_i)| < 1.$$

Hence

$$|f(z)| < |f(c_i)| + 1 \le K.$$

Let $M = \sup_S |f|$, and suppose, for a contradiction, that $|f(z)| < M$ for all z in S. It follows that the function $g : S \to \mathbb{R}$ given by

$$g(z) = 1/(M - |f(z)|)$$

being continuous, is bounded in S. On the other hand, for all $K > 0$ there exists z in S such that $M - |f(z)| < 1/K$ (for otherwise a smaller bound would be possible). Thus $|g(z)| = g(z) > K$, and so g is not bounded. From this contradiction we gain the required result, that the function $|f|$ attains its supremum within S.   □

The following result is an easy consequence:

Corollary 5.4

Let S be a closed bounded set, and let f, with domain containing S, be continuous and non-zero throughout S. Then $\inf\{|f(z)| : x \in S\} > 0$.

Proof

From the hypotheses we see that $1/f$ is continuous throughout S. Hence there exists $M > 0$ such that $\sup\{|1/f(z)| : z \in S\} = M$. It follows that $\inf\{|f(z)| : x \in S\} = 1/M > 0$. $\qquad\Box$

EXERCISES

5.1 Show that both "closed" and "bounded" are required in Theorem 5.3.

5.2 Parametric Representation

It will be convenient in this section to define curves by means of a **parametric representation**. That is, a **curve**, or **path**, $\mathcal{C}$ is defined as

$$\mathcal{C} = \{(r_1(t), r_2(t)) : t \in [a, b]\},$$

where $[a, b]$ is an interval, and r_1, r_2 are real continuous functions with domain $[a, b]$.

This has some advantages over the standard approach

$$\mathcal{C} = \{(x, f(x)) : x \in [a, b]\},$$

for there are no problems when the curve becomes vertical, or crosses itself:

The other advantage is that the definition imposes an **orientation**, which is the direction of travel of a point on the curve as t increases from a to b.

We shall find it useful to use vector notation and to write

$$\mathcal{C} = \{\mathbf{r}(t) : t \in [a, b]\}$$

where $\mathbf{r}(t)$ is the vector $\big(r_1(t), r_2(t)\big)$.

If $\mathbf{r}(a) = \mathbf{r}(b)$ we say that $\mathcal{C}$ is a **closed** curve. If $a \leq t < t' \leq b$ and $|t' - t| < b - a$ implies that $\mathbf{r}(t) \neq \mathbf{r}(t')$, we say that $\mathcal{C}$ is a **simple** curve. Visually, a simple curve does not cross itself.

Some examples will help.

Example 5.5

If $\mathbf{r}(t) = (\cos t, \sin t)$ $(t \in [0, 2\pi])$, then $\mathcal{C}$ is a simple closed curve. The curve, a circle of radius 1, begins and ends at the point $(1, 0)$, and the orientation is anticlockwise.

Example 5.6

Let $\mathbf{r}(t) = (t^2, t)$ $\big(t \in [-1, 1]\big)$. The curve, a parabola, is simple but not closed. It begins at $A(1, -1)$ and ends at $B(1, 1)$, and the orientation is as shown.

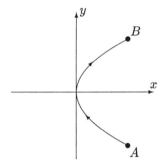

Example 5.7

Let $\mathbf{r}(t) = (\cos t \cos 2t, \sin t \cos 2t)$ $\big(t \in [0, 2\pi]\big)$. This is a closed curve, but is not simple, since

$$\mathbf{r}(\tfrac{\pi}{4}) = \mathbf{r}(\tfrac{3\pi}{4}) = \mathbf{r}(\tfrac{5\pi}{4}) = \mathbf{r}(\tfrac{7\pi}{4}) = (0, 0).$$

As t increases from 0 to 2π the point $\mathbf{r}(t)$ follows a smooth path from A to O

to B to O to C to O to D to O and back to A:

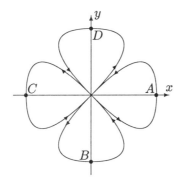

Consider a curve

$$\mathcal{C} = \{\mathbf{r}(t) \,:\, t \in [a,b]\}\,,$$

and let $D = \{a = t_0, t_1, \ldots, t_n = b\}$ be a dissection of $[a,b]$, with

$$t_0 < t_1 < \cdots < t_n\,.$$

Each t_i in D corresponds to a point $P_i = \mathbf{r}(t_i)$ on the curve $\mathcal{C}$, and it is reasonable to estimate the length of curve $\mathcal{C}$ between the point $A = P_0$ and $B = P_n$ as

$$\mathcal{L}(\mathcal{C}, D) = |P_0 P_1| + |P_1 P_2| + \cdots + |P_{n-1} P_n|\,. \tag{5.1}$$

In analytic terms, this becomes

$$\mathcal{L}(\mathcal{C}, D) = \sum_{i=1}^{n} \|\mathbf{r}(t_i) - \mathbf{r}(t_{i-1})\|\,, \tag{5.2}$$

where, for a two-dimensional vector $\mathbf{v} = (v_1, v_2)$, we define $\|\mathbf{v}\|$, the **norm** of $\mathbf{v}$, to be $\sqrt{v_1^2 + v_2^2}$.

It is clear that if we refine the dissection D by adding extra points then $\mathcal{L}(\mathcal{C}, D)$ increases: if Q is a point between P_{i-1} and P_i, then, by the triangle inequality, the combined length of segments $P_{i-1}Q$ and QP_i is not less than the length of the segment $P_{i-1}P_i$.

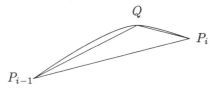

Let $\mathcal{D}$ be the set of all dissections of $[a,b]$. If $\{\mathcal{L}(\mathcal{C}, D) \,:\, D \in \mathcal{D}\}$ is bounded above, we say that the curve $\mathcal{C}$ is **rectifiable**, and we define its **length** $\Lambda(\mathcal{C})$ by

$$\Lambda(\mathcal{C}) = \sup \{\mathcal{L}(\mathcal{C}, D) \,:\, D \in \mathcal{D}\}\,.$$

Not every curve is rectifiable:

Example 5.8

Let $C = \{(t, r_2(t)) : t \in [0,1]\}$, where

$$r_2(t) = \begin{cases} t\sin(1/t) & \text{if } t \neq 0 \\ 0 & \text{if } t = 0. \end{cases}$$

Show that C is not rectifiable.

Solution

For $n = 1, 2, 3, \ldots$, let

$$D_n = \left\{ 0, \frac{2}{n\pi}, \frac{2}{(n-1)\pi}, \ldots, \frac{2}{2\pi}, \frac{2}{\pi}, 1 \right\}.$$

Observe that

$$r_2\left(\frac{2}{k\pi}\right) = \frac{2}{k\pi}\sin\left(\frac{k\pi}{2}\right) = \begin{cases} 0 & \text{if } k \text{ is even} \\ \pm 2/k\pi & \text{if } k \text{ is odd.} \end{cases}$$

Hence, if k is even,

$$\left\| \mathbf{r}\left(\frac{2}{k\pi}\right) - \mathbf{r}\left(\frac{2}{(k+1)\pi}\right) \right\| = \left\| \left(\frac{2}{k\pi}, 0\right) - \left(\frac{2}{(k+1)\pi}, \pm\frac{2}{(k+1)\pi}\right) \right\|$$

$$> \frac{2}{(k+1)\pi},$$

and if k is odd we can similarly show that

$$\left\| \mathbf{r}\left(\frac{2}{k\pi}\right) - \mathbf{r}\left(\frac{2}{(k+1)\pi}\right) \right\| > \frac{2}{k\pi} > \frac{2}{(k+1)\pi}.$$

It follows that

$$\mathcal{L}(C, D_n) > \frac{2}{\pi}\left(\frac{1}{2} + \frac{1}{3} + \cdots \frac{1}{n}\right),$$

and from the divergence of the harmonic series we see that there is no upper bound on the set $\{\mathcal{L}(C, D) : D \in \mathcal{D}\}$. □

The following theorem, whose proof can be found in [9, Theorem 8.5], identifies a wide class of rectifiable curves, and gives a formula for their lengths:

Theorem 5.9

Let $C = \{\mathbf{r}(t) : t \in [a, b]\}$, where $\mathbf{r}(t) = (r_1(t), r_2(t))$, and suppose that r_1, r_2 are differentiable and r_1', r_2' are continuous on $[a, b]$. Then C is rectifiable, and the length $\Lambda(C)$ of C is given by

$$\Lambda(C) = \int_a^b \|\mathbf{r}'(t)\| \, dt. \tag{5.3}$$

□

Here $\mathbf{r}'(t) = \big(r_1'(t), r_2'(t)\big)$, and so we have the alternative formula

$$\Lambda(\mathcal{C}) = \int_a^b \sqrt{\big(r_1'(t)\big)^2 + \big(r_2'(t)\big)^2}\, dt\,. \tag{5.4}$$

We can easily "translate" a pair $\big(\alpha(t), \beta(t)\big)$ of real continuous functions defined on an interval $[a, b]$ into a continuous function $\gamma : [a, b] \to \mathbb{C}$, where

$$\gamma(t) = \alpha(t) + i\beta(t) \quad \big(t \in [a, b]\big)\,.$$

Thus, in Example 5.5, $\gamma(t) = e^{it}$, and in Example 5.7, $\gamma(t) = e^{it}\cos 2t$. The image of γ is the curve

$$\gamma^* = \{\gamma(t) : t \in [a, b]\}\,.$$

Observe that $\|\mathbf{r}(t)\|$ translates to $|\gamma(t)|$, so that (5.3) becomes

$$\Lambda(\gamma^*) = \int_a^b |\gamma'(t)|\, dt\,. \tag{5.5}$$

The formula applies if γ is **smooth**, that is to say, if γ has a continuous derivative in $[a, b]$.

We shall not always be meticulous about preserving the distinction between the function γ and the associated curve γ^*.

Example 5.10

Determine the length of the circumference of the circle $\{re^{it} : 0 \le t \le 2\pi\}$.

Solution

Here $|\gamma'(t)| = |ire^{it}| = r$, and so, with some relief, we see that

$$\Lambda = \int_0^{2\pi} r\, dt = 2\pi r\,.$$

$\square$

Example 5.11

Find the length of γ^*, where

$$\gamma(t) = t - ie^{-it} \quad (0 \le t \le 2\pi)\,.$$

Solution

Since

$$\gamma'(t) = 1 - e^{-it} = (1 - \cos t) + i \sin t,$$

it follows that

$$|\gamma'(t)|^2 = (1 - \cos t)^2 + \sin^2 t = 2 - 2 \cos t = 4 \sin^2 \frac{t}{2}.$$

Hence, since $\sin(t/2)$ is non-negative throughout the interval $[0, 2\pi]$,

$$|\gamma'(t)| = 2 \sin \frac{t}{2}.$$

Thus

$$\Lambda(\gamma^*) = \int_0^{2\pi} 2 \sin \frac{t}{2}\, dt = \left[-4 \cos \frac{t}{2} \right]_0^{2\pi} = 8.$$

$\square$

Remark 5.12

The curve γ^* is called a **cycloid**, and looks like this:

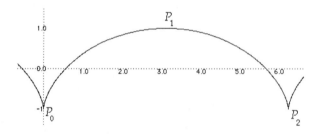

Figure 5.1. A cycloid

It is the path of a point on the circumference of a wheel of radius 1 rolling along the line $y = -1$ and making one complete rotation. The points P_0, P_1 and P_2 correspond respectively to the values 0, π and 2π of t.

EXERCISES

5.2. For any two distinct complex numbers, the line segment from c to d can be parametrised by

$$\gamma(t) = (1 - t)c + td \quad (0 \le t \le 1).$$

By using (5.5), verify that the length of the line segment is what it ought to be.

5.3. Sketch the curves

   a) $\{(a\cos t, b\sin t) : 0 \le t \le 2\pi\}$ $(a, b > 0)$;

   b) $\{(a\cosh t, b\sinh t) : t \in [0, \infty)\}$ $(a, b > 0)$;

   c) $\{(at, a/t) : t \in (0, \infty)\}$ $(a > 0)$.

5.4. Sketch the curve $\{te^{it} : t \in [0, 2\pi]\}$, and determine its length.

5.5. Let $a, b \in \mathbb{R}$, with $a < b$. Determine the length of $\Lambda(a, b)$ of the curve $\{e^{t+it} : t \in [a, b]\}$. Determine $\lim_{a \to -\infty} \Lambda(a, b)$.

5.3 Integration

We aim to define the integral of a complex function along a curve in the complex plane: *a cocplex line cocegal*

Let $\gamma : [a, b] \to \mathbb{C}$ be smooth, and let f be a "suitable" complex function whose domain includes the curve γ^*. We define

$$\int_\gamma f(z)\, dz = \int_a^b f\big(\gamma(t)\big)\gamma'(t)\, dt. \tag{5.6}$$

This does require a bit of explanation. First, if we define $g : [a, b] \to \mathbb{C}$ by

$$g(t) = f\big(\gamma(t)\big)\gamma'(t),$$

then $g(t) = u(t) + iv(t)$, where u, v are functions from $[a, b]$ to $\mathbb{R}$, and we define $\int_a^b g(t)\, dt$ in the obvious way by

$$\int_a^b g(t)\, dt = \int_a^b u(t)\, dt + i \int_a^b v(t)\, dt.$$

Secondly, f is "suitable" if and only if the right hand side of (5.6) is defined, that is, if and only if $f\big(\gamma(t)\big)\gamma'(t)$ is **integrable**. The reader who is familiar with some version of formal integration theory will know what this means, but for our purposes it is sufficient to know that every continuous function f is suitable.

We refer to $\int_\gamma f(z)\, dz$ as the **integral of f along** γ. If γ is a closed curve, we call it **the integral round** γ.

The following easy result has, as we shall see in Chapter 8, far-reaching consequences.

Theorem 5.13

Let $\gamma(t) = e^{it}$ $(0 \le t \le 2\pi)$, so that γ^* is the unit circle (with centre 0 and radius 1), and let n be an integer. Then

$$\int_\gamma z^n \, dz = \begin{cases} 2\pi i & \text{if } n = -1 \\ 0 & \text{otherwise.} \end{cases}$$

Proof

By (5.6),

$$\int_\gamma z^n \, dz = \int_0^{2\pi} r^n e^{nit} i e^{it} \, dt = i \int_0^{2\pi} r^n e^{(n+1)it} \, dt \, .$$

If $n = -1$ this becomes

$$i \int_0^{2\pi} dt = 2\pi i \, .$$

Otherwise

$$\int_\gamma z^n \, dz = i \int_0^{2\pi} r^n \big[\cos(n+1)t + i \sin(n+1)t \big] \, dt$$

$$= i r^n \left(\left[\frac{\sin(n+1)t}{n+1} - i \frac{\cos(n+1)t}{n+1} \right]_0^{2\pi} \right) = 0 \, .$$

$\square$

Remark 5.14

We shall see shortly that it is legitimate to shorten the argument by writing

$$\int_0^{2\pi} r^n e^{(n+1)it} \, dt = \left[\frac{r^n}{(n+1)i} e^{(n+1)it} \right]_0^{2\pi} = 0 \, .$$

Remark 5.15

Although we write $\int_\gamma$, the integral does not depend on the particular (smooth increasing) parametrisation of the contour γ^*. Thus, for example, if in Theorem 5.13 we were to parametrise the circle by $\delta(t) = e^{2it}$ $(0 \le t \le \pi)$ the value of the integral would not change.

In Theorem 5.13 the parametrisation $\gamma(t) = e^{it}$ implied that the closed curve was traversed in the positive (anti-clockwise) direction. If it is traversed in the other direction we must take $\gamma(t) = e^{-it}$. If $n \neq -1$ this makes no difference to the answer, but if $n = -1$ we obtain

$$\int_\gamma z^{-1}\,dz = \int_0^{2\pi} e^{it}(-ie^{-it})\,dt = -2\pi i\,.$$

In general, if $\gamma : [a,b] \to \mathbb{C}$ is a curve, we define $\overleftarrow{\gamma}$, the same path but with the opposite orientation, by

$$(\overleftarrow{\gamma})(t) = \gamma(a+b-t) \quad (t \in [a,b])\,. \tag{5.7}$$

Then we have

$$\int_{\overleftarrow{\gamma}} f(z)\,dz = -\int_\gamma f(z)\,dz\,. \tag{5.8}$$

To see this, observe that

$$\begin{aligned}
\int_{\overleftarrow{\gamma}} f(z)\,dz &= \int_a^b f\big(\overleftarrow{\gamma}(t)\big)(\overleftarrow{\gamma})'(t)\,dt \\
&= \int_a^b f\big(\gamma(a+b-t)\big)\big(-\gamma'(a+b-t)\big)\,dt \\
&= \int_b^a f\big(\gamma(u)\big)\big(-\gamma'(u)\big)(-du)\,, \text{ where } u = a+b-t, \\
&= -\int_a^b f\big(\gamma(u)\big)\gamma'(u)\,du = -\int_\gamma f(z)\,dz\,.
\end{aligned}$$

At this point it is important to recall some of the standard properties of integrals. For real numbers $a < b < c$, real functions f and g, and a constant k,

$$\int_a^b (f \pm g) = \int_a^b f \pm \int_a^b g\,, \tag{5.9}$$

$$\int_a^b kf = k \int_a^b f\,, \tag{5.10}$$

$$\int_a^b f + \int_b^c f = \int_a^c f\,. \tag{5.11}$$

We define $\int_a^a f = 0$, and if $a > b$ we define

$$\int_a^b f = -\int_b^a f\,.$$

With these conventions, (5.11) holds for arbitrary real numbers a, b and c. These formulae easily extend to functions $f, g : \mathbb{R} \to \mathbb{C}$, and to the case where the constant k is complex.

The requirement in (5.6) that γ be a smooth function is inconveniently restrictive, and there is no great difficulty in extending the definition to the case where γ is **piecewise smooth**. Geometrically, the curve consists of finitely many smooth segments:

Analytically, $\gamma : [a, b] \to \mathbb{C}$ is **piecewise smooth** if there are real numbers

$$a = c_0 < c_1 < \cdots < c_m = b$$

and smooth functions $\gamma_i : [c_{i-1}, c_i] \to \mathbb{C}$ $(i = 1, \ldots, m)$ such that

$$\gamma_i(c_i) = \gamma_{i+1}(c_i) \quad (i = 1, \ldots, m-1).$$

Then

$$\int_\gamma f(z)\,dz = \sum_{i=1}^m \left(\int_{\gamma_i} f(z)\,dz \right). \tag{5.12}$$

In practice we proceed in a slightly different way.

Example 5.16

Let $\gamma = \sigma(0, R)$ be the closed semicircle shown.

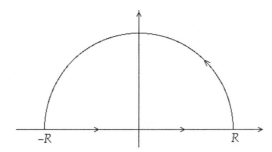

Determine $\int_\gamma z^2\,dz$.

Solution

The curve is in two sections, first the line segment γ_1 from $(-R, 0)$ to $(R, 0)$, then the semicircular arc γ_2 (in the positive direction. The easiest way to parametrise γ_1 is by

$$\gamma_1(t) = t + i0 \quad \left(t \in [-R, R] \right),$$

and so

$$\int_{\gamma_1} z^2 \, dz = \int_{-R}^{R} t^2 \, dt = \frac{2R^3}{3}.$$

Again, it is natural to take

$$\gamma_2(t) = Re^{it} \quad (t \in [0, \pi]),$$

and so, from (5.6)

$$\int_{\gamma_2} z^2 \, dz = \int_0^{\pi} R^2 e^{2it} \cdot Rie^{it} = \left[\frac{1}{3} R^3 e^{3it}\right]_0^{\pi} = -\frac{2R^3}{3}.$$

Hence

$$\int_{\gamma} z^2 \, dz = \int_{\gamma_1} z^2 \, dz + \int_{\gamma_2} z^2 \, dz = 0.$$

$\square$

Remark 5.17

The procedure adopted is not quite in accordance with (5.12). It is, however, always possible to carry out the "official" procedure by re-parametrising γ_1 and γ_2 so that the end of the interval domain of γ_1 coincides with the beginning of the interval domain of γ_2. In our example we could define

$$\gamma_1(t) = 4Rt - R \quad \left(t \in [0, \tfrac{1}{2}]\right), \qquad \gamma_2(t) = Re^{i\pi(2t-1)} \quad \left(t \in [\tfrac{1}{2}, 1]\right).$$

The answer is the same as before, since the changes simply amount to making a substitution in the integral:

$$\int_{\gamma_1} z^2 \, dz = \int_0^{1/2} (4Rt - R)^2 4R \, dt = \left[\frac{1}{3}(4Rt - R)^3\right]_0^{1/2} = \frac{2R^3}{3};$$

$$\int_{\gamma_2} z^2 \, dz = \int_{1/2}^1 R^2 e^{2i\pi(2t-1)} 2i\pi Re^{i\pi(2t-1)} \, dt$$

$$= R^3 \int_0^{\pi} ie^{3iu} \, du \quad (\text{where } u = 2\pi(2t-1))$$

$$= \frac{R^3}{3} \left[e^{3iu}\right]_0^{\pi} = -\frac{2R^3}{3}.$$

All we achieve by doing it this way is more likelihood of error because of the extra technical difficulty involved.

We have already dared to suppose that some of the rules of real variable calculus might apply to complex-valued functions. In solving Example 5.16 we wrote

$$\int_0^\pi e^{3it}\, dt = \left[\frac{1}{3i}e^{3it}\right]_0^\pi = -\frac{2}{3i} = \frac{2i}{3}\,.$$

This certainly works, for if we do it the hard way we have

$$\int_0^\pi e^{3it}\, dt = \int_0^\pi (\cos 3t + i\sin 3t)\, dt = \frac{1}{3}\left[\sin 3t - i\cos 3t\right]_0^\pi = \frac{2i}{3}\,.$$

We might suspect that there is a theorem lurking in the shadows, and we would be right:

Theorem 5.18

Let $f : [a, b] \to \mathbb{C}$ be continuous, and let

$$F(x) = \int_a^x f(t)\, dt \quad (x \in [a, b])\,.$$

Then $F'(x) = f(x)$ for all x in $[a, b]$. If $\Theta : [a, b] \to \mathbb{C}$ is any function such that $\Theta' = f$, then

$$\int_a^b f(t)\, dt = \Theta(b) - \Theta(a)\,.$$

Proof

The proof is entirely routine, and depends on the separation of real and imaginary parts. Suppose that $\operatorname{Re} f = g$, $\operatorname{Im} f = h$. Then

$$F(x) = \int_a^x [g(t) + ih(t)]\, dt = \int_a^x g(t)\, dt + i\int_a^x h(t)\, dt$$
$$= G(x) + iH(x) \quad \text{(say)}.$$

Hence, by the Fundamental Theorem of Calculus,

$$F'(x) = G'(x) + iH'(x) = g(x) + ih(x) = f(x)\,.$$

Suppose now that $\Theta' = f$. Writing $\Theta = \Phi + i\Psi$ in the usual way, we see that

$$\Phi' = G' = g, \quad \Psi' = H' = h\,,$$

and so, for some constants C, C', and for all x in $[a, b]$,

$$G(x) = \Phi(x) + C, \quad H(x) = \Psi(x) + C'\,.$$

Putting $x = a$ gives $C = -\Phi(a)$, $C' = -\Psi(a)$, and so

$$\int_a^b f(t)\,dt = G(b) + iH(b) = \big(\Phi(b) - \Phi(a)\big) + i\big(\Psi(b) - \Psi(a)\big)$$
$$= \big(\Phi(b) + i\Psi(b)\big) - \big(\Phi(a) + i\Psi(a)\big) = \Theta(b) - \Theta(a)\,.$$

$\square$

As a consequence of this result we have the following theorem:

Theorem 5.19

Let $\gamma : [a, b] \to \mathbb{C}$ be piecewise smooth. Let F be a complex function defined on an open set containing γ^*, and suppose that $F'(z)$ exists and is continuous at each point of γ^*. Then

$$\int_\gamma F'(z)\,dz = F\big(\gamma(b)\big) - F\big(\gamma(a)\big)\,.$$

In particular, if γ is closed, then

$$\int_\gamma F'(z)\,dz = 0\,.$$

Proof

Suppose first that γ is smooth. The assumptions imply that $F \circ \gamma$ is differentiable on $[a, b]$. Since $(F \circ \gamma)'(t) = F'\big(\gamma(t)\big)\gamma'(t)$, it follows from Theorem 5.18 that

$$\int_\gamma F'(z)\,dz = \int_a^b (F \circ \gamma)'(t)\,dt = F\big(\gamma(b)\big) - F\big(\gamma(a)\big)\,. \qquad (5.13)$$

Now suppose that γ is piecewise smooth. That is, suppose that there are real numbers $a = c_0 < c_1 < \cdots < c_m = b$ and smooth functions $\gamma_i : [c_{i-1}, c_i] \to \mathbb{C}$ $(i = 1, \ldots, m)$ such that

$$\gamma_i(c_i) = \gamma_{i+1}(c_i) \quad (i = 1, \ldots, m-1)\,, \qquad (5.14)$$

and, for each i, γ coincides with γ_i in the interval $[c_{i-1}, c_i]$. Then (5.13) applies to each of the functions γ_i and so

$$\int_\gamma F'(z)\,dz = \sum_{i=1}^m \int_{\gamma_i} F'(z)\,dz$$
$$= \big[F\big(\gamma_1(c_1)\big) - F\big(\gamma_1(a)\big)\big] + \big[F\big(\gamma_2(c_2)\big) - F\big(\gamma_2(c_1)\big)\big] + \cdots$$

$$\cdots + \left[F\big(\gamma_m(b)\big) - F\big(\gamma_m(c_{m-1})\big) \right]$$
$$= F\big(\gamma_m(b)\big) - F\big(\gamma_1(a)\big) \quad \text{(by (5.14))}$$
$$= F\big(\gamma(b)\big) - F\big(\gamma(a)\big) \,.$$

If γ is closed, then $\gamma(b) = \gamma(a)$, and so $\int_\gamma F'(z)\, dz = 0$. $\qquad\square$

Two easy cases are worth recording formally as a corollary:

Corollary 5.20

Let γ be piecewise smooth and closed. Then

$$\int_\gamma 1\, dz = \int_\gamma z\, dz = 0 \,.$$

Proof

In the theorem, take first $F(z) = z$, then $F(z) = z^2/2$. $\qquad\square$

Example 5.21

Let γ^* be the top half of the ellipse

$$\frac{x^2}{a^2} + \frac{y^2}{b^2} = 1 \,,$$

traversed in the positive (counterclockwise) direction. Determine

$$\int_\gamma \cos z\, dz \,.$$

Solution

The ellipse meets the x-axis at the points $(a, 0)$ and $(-a, 0)$. By Theorem 5.19, we do not need to find a parametrisation of γ:

$$\int_\gamma \cos z\, dz = \left[\sin z \right]_a^{-a} = -2 \sin a \,.$$

$\qquad\square$

Remark 5.22

A closed piecewise smooth curve γ can be quite a complicated object, but Theorem 5.19 assures us that, for suitable functions F, the integral of F' round γ has the value 0.

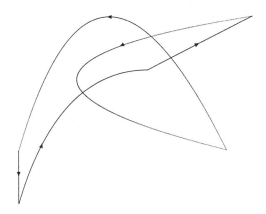

For the most part, however, we shall be interested in **simple** curves, with no crossings.

In its most general aspect, a contour is a simple closed curve. Here we shall make a more restrictive definition, for we shall not be considering anything more general: a **contour** in $\mathbb{C}$ is a *piecewise smooth, simple, closed curve*. This definition is certainly general enough to cover all the significant applications in a book at this level. Unless we specify otherwise, we shall assume that contours are traversed in the positive (anti-clockwise) direction.

Let $\gamma : [a, b] \to \mathbb{C}$ be a piecewise smooth function, determining a simple closed curve γ^*.

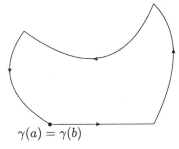

$$\gamma(a) = \gamma(b)$$

We shall regard it as geometrically obvious that the complement of the set $\gamma*$ is the disjoint union of two open sets $\mathrm{I}(\gamma)$, called the **interior** of γ^*, and $\mathrm{E}(\gamma)$, called the **exterior** of γ^*. The interior is bounded and the exterior unbounded. In its general form this is the **Jordan Curve Theorem**, and a proper proof, not appealing to geometric intuition, is difficult. It appears in Jordan's[3] *Cours d'Analyse* of 1887. I am *not* suggesting that you read this: in a much more ambitious book than this, Ahlfors [1] remarks that "no proof is included of the Jordan Curve Theorem, which, to the author's knowledge, is never needed in function theory."

---

[3] Marie Ennemond Camille Jordan, 1838–1922.

EXERCISES

5.6. Let $\gamma(t) = z - a - t \ (0 \le t \le h)$. Show that

$$\int_\gamma \frac{d\zeta}{\zeta^{n+1}} = \frac{1}{n} \left(\frac{1}{(z-a-h)^n} - \frac{1}{(z-a)^n} \right).$$

5.7. Evaluate $\int_\gamma f(z)\, dz$, where

a) $f(z) = \operatorname{Re} z$, $\gamma(t) = t^2 + it$, $t \in [0, 1]$;

b) $f(z) = z^2$, $\gamma(t) = e^{it}$, $t \in [0, \pi]$;

c) $f(z) = 1/z$, $\gamma(t) = e^{it}$, $t \in [0, 6\pi]$;

d) $f(z) = \cos z$, where γ^* is the straight line from $-\pi - i\pi$ to $\pi - i\pi$ followed by the straight line from $\pi - i\pi$ to $\pi + i\pi$.

5.4 Estimation

In analysis, both real and complex, it is important to be able to estimate a quantity, especially to put a bound on its value. The results of this section will prove to be powerful tools as the theory develops.

Theorem 5.23

Let $g : [a, b] \to \mathbb{C}$ be continuous. Then

$$\left| \int_a^b g(t)\, dt \right| \le \int_a^b |g(t)|\, dt.$$

Proof

We know (see [9, Theorem 5.15 (iv)]) that the inequality holds for every integrable *real* function g,

$$\left| \int_a^b g(t)\, dt \right| \le \int_a^b |g(t)|\, dt.$$

To show that it holds also for complex-valued functions, observe first that, for some θ,

$$\left| \int_a^b g(t)\, dt \right| = e^{i\theta} \int_a^b g(t)\, dt = \int_a^b e^{i\theta} g(t)\, dt$$

$$= \int_a^b \mathrm{Re}\big[e^{i\theta} g(t)\big]\, dt + i \int_a^b \mathrm{Im}\big[e^{i\theta} g(t)\big]\, dt\,.$$

Since the imaginary part of the left-hand side is zero, we deduce that

$$\left| \int_a^b g(t)\, dt \right| = \int_a^b \mathrm{Re}\big[e^{i\theta} g(t)\big]\, dt$$

$$\leq \int_a^b \big| e^{i\theta} g(t) \big|\, dt \quad \text{(by Theorem 2.1)}$$

$$= \int_a^b |g(t)|\, dt\,.$$

$\square$

As a consequence, we have

Theorem 5.24

Let $\gamma : [a, b] \to \mathbb{C}$ be piecewise smooth, and let f be a continuous complex function with the property that $|f(z)| \leq M$ for all z in γ^*. Then

$$\left| \int_\gamma f(z)\, dz \right| \leq Ml\,,$$

where $l = \int_a^b |\gamma'(t)|\, dt$ is the length of γ^*.

Proof

By Theorem 5.23,

$$\left| \int_\gamma f(z)\, dz \right| = \left| \int_a^b f\big(\gamma(t)\big) \gamma'(t)\, dt \right| \leq \int_a^b \big| f\big(\gamma(t)\big) \big| |\gamma'(t)|\, dt$$

$$\leq M \int_a^b |\gamma'(t)|\, dt = Ml\,.$$

$\square$

Theorem 5.24 has many applications. First, by a **convex** contour we mean a contour γ with the property that, for all a, b in $I(\gamma)$, the line segment connecting a and b lies wholly in $I(\gamma)$.

Theorem 5.25

Let γ be a convex contour, and suppose that $\int_\tau f(z)\,dz = 0$ for every triangular contour τ within $I(\gamma)$. Then there exists a function F, holomorphic in $I(\gamma)$, such that $F'(z) = f(z)$ for all z in $I(\gamma)$.

Proof

Let $z \in I(\gamma)$. Since $I(\gamma)$ is open, there exists $\delta > 0$ such that $N(z,\delta) \subset I(\gamma)$. If we choose h in $\mathbb{C}$ so that $|h| < \delta$, we can be sure that $z + h \in I(\gamma)$. Let a be an arbitrary fixed point in $I(\gamma)$. Then, by the convexity of $I(\gamma)$, the entire triangle with vertices at a, z and $z + h$ lies within $I(\gamma)$.

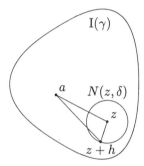

Our hypothesis concerning f implies that

$$\int_{[a,z]} f(w)\,dw + \int_{[z,z+h]} f(w)\,dw - \int_{[a,z+h]} f(w)\,dw = 0;$$

hence, if we define $F(z)$ as $\int_{[a,z]} f(w)\,dw$, we have that

$$F(z+h) - F(z) = \int_{[z,z+h]} f(w)\,dw.$$

We know that, for a constant k, $\int_{[z,z+h]} k\,dw = kh$, and so in particular

$$\int_{[z,z+h]} f(z)\,dw = hf(z).$$

By Theorem 5.3, the function f, being continuous, is bounded within $I(\gamma)$, and so, by Theorem 5.24,

$$\left| \frac{F(z+h) - F(z)}{h} - f(z) \right| = \frac{1}{|h|} \left| \int_{[z,z+h]} \big(f(w) - f(z)\big)\,dw \right|$$

$$\leq \frac{1}{|h|} \cdot |h| \sup_{w \in [z,z+h]} |f(w) - f(z)|.$$

Since this, again by the continuity of f, tends to zero as $h \to 0$, we deduce that F' exists and that $F' = f$. $\qquad\square$

In view of Theorem 5.19, this theorem has an immediate consequence:

Corollary 5.26

If γ is convex, and if $\int_\tau f(z)\,dz = 0$ for every triangular contour τ within $I(\gamma)$, then $\int_\gamma f(z)\,dz = 0$.

Example 5.27

Let $f(z) = 1/(z^3 + 1)$ and let $\gamma(t) = Re^{it}$ ($0 \le t \le \pi$). Show that

$$\lim_{R \to \infty} \left| \int_\gamma f(z)\,dz \right| = 0 \,.$$

Solution

For each point on γ^*,

$$|f(z)| = \left| \frac{1}{R^3 e^{3it} + 1} \right| = \frac{1}{|R^3 e^{3it} + 1|} \,.$$

Now, by Theorem 2.1,

$$|R^3 e^{3it} + 1| \ge |R^3 e^{3it}| - 1 = R^3 - 1 \,,$$

and the length of γ^* is $R\pi$. Hence

$$0 \le \left| \int_\gamma f(z)\,dz \right| \le \frac{R\pi}{R^3 - 1} \,,$$

which clearly tends to zero as R tends to infinity. $\qquad\square$

Example 5.28

Let $f(z) = 1/(z + \frac{1}{2})$ and let $\gamma(t) = e^{it}$ ($0 \le t \le \pi$). Show that

$$\left| \int_\gamma f(z)\,dz \right| \le 2\pi \,.$$

Solution

Here

$$|f(z)| = \frac{1}{|z + \frac{1}{2}|} \le \frac{1}{|z| - \frac{1}{2}} \le 2$$

for all z on γ^*, and the length of the curve is π. The required inequality follows immediately.   □

EXERCISES

5.8. Let $\gamma(t) = (1 - t)i + t$ $(0 \le t \le 1)$, so that γ^* is the straight line from i to 1. Show that, for all z on γ,

$$|z^4| \ge \frac{1}{4},$$

and deduce that, if

$$I = \int_\gamma \frac{dz}{z^4},$$

then $|I| \le 4\sqrt{2}$. What is the true value of $|I|$?

5.9. Let

$$f(z) = \frac{z^3 - 4z + 1}{(z^2 + 5)(z^3 - 3)},$$

and let $\gamma(t) = Re^{it}$ $(0 \le t \le \pi)$. Show that

$$\left| \int_\gamma f(z)\, dz \right| \le \frac{\pi R(R^3 + 4R + 1)}{(R^2 - 5)(R^3 - 3)}.$$

5.10. Show that, if u and v are real,

$$|\sin(u + iv)| \le \cosh 2v.$$

5.11. Let γ^* be as shown

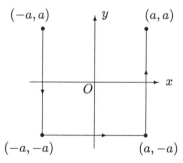

and let $f(z) = \sin(z^2)$. Show that

$$\left| \int_\gamma f(z)\, dz \right| \le 6a \cosh(4a^2).$$

5.5 Uniform Convergence

In Theorem 4.17 we established that the process of differentiating a power series (within the circle of convergence) term by term is valid. It was useful to have that result at an early stage, but in fact it is an instance of a general result in a theory whose key notion is that of **uniform convergence**. It is convenient to finish this chapter with a brief account of the main ideas. We include only results that will be used later.

Let f be a bounded complex function whose domain is a subset S of $\mathbb{C}$. We define $\|f\|$, the **norm** of f, by

$$\|f\| = \sup_{z \in S} |f(z)| . \tag{5.15}$$

It is clear that $\|f\| \geq 0$, and that $\|f\| = 0$ only if $f = 0$. Less obviously, we also have the **triangle inequality**: if f and g have the same domain, then

$$\|f + g\| \leq \|f\| + \|g\| . \tag{5.16}$$

To see this, observe that, for all z in the common domain S of f and g,

$$|f(z) + g(z)| \leq |f(z)| + |g(z)| \leq \|f\| + \|g\| .$$

Hence

$$\|f + g\| = \sup_{z \in S} |f(z) + g(z)| \leq \|f\| + \|g\| .$$

Let f_n be a sequence of complex functions whose domain is a subset S of $\mathbb{C}$. We say that (f_n) **tends uniformly in S to f**, or **has uniform limit f**, or is **uniformly convergent to f**, if, for every $\epsilon > 0$ there exists a positive integer N such that $\|f - f_n\| < \epsilon$ for all $n > N$. This certainly implies that, for each z in S, the sequence $(f_n(z))$ tends to $f(z)$, but the converse implication may be false.

Example 5.29

Let

$$f_n(z) = \frac{1 - z^n}{1 - z} \quad \left(z \in N(0, 1)\right) .$$

Show that (f_n) converges pointwise, but not uniformly, to f, where $f(z) = 1/(1 - z)$.

Solution

For each z in $N(0, 1)$,

$$\lim_{n \to \infty} f_n(z) = \frac{1}{1 - z} .$$

(We say that (f_n) tends **pointwise** to f, where $f(z) = 1/(1-z)$.) On the other hand, for each fixed n, $f - f_n$ is not even bounded in the set $N(0,1)$, since, as $z \to 1$,

$$\frac{1}{f(z) - f_n(z)} = \frac{1-z}{z^n} \to \frac{0}{1} = 0.$$

The convergence is not uniform.   □

Theorem 5.30

Let f_n be a sequence of functions with common domain S, converging uniformly in S to a function f. Let $a \in S$. If each f_n is continuous at a, then so is f.

Proof

Let $\epsilon > 0$ be given. There exists N such that $\|f - f_n\| < \epsilon/3$ for all $n > N$, and there exists $\delta > 0$ such that $|f_{N+1}(z) - f_{N+1}(a)| < \epsilon/3$ for all z such that $|z - a| < \delta$. Hence, for all such z,

$$
\begin{aligned}
|f(z) - f(a)| &= |(f(z) - f_{N+1}(z)) + (f_{N+1}(z) - f_{N+1}(a)) + (f_{N+1}(a) - f(a))| \\
&\leq |f(z) - f_{N+1}(z)| + |f_{N+1}(z) - f_{N+1}(a)| + |f(a) - f_{N+1}(a)| \\
&\leq \|f - f_{N+1}\| + |f_{N+1}(z) - f_{N+1}(a)| + \|f - f_{N+1}\| \\
&< \epsilon,
\end{aligned}
$$

and so f is continuous at a.   □

The idea of uniform convergence applies also to *series* of functions. Given a sequence (f_n) of functions with common domain S, we define the function F_n

$$F_n(z) = \sum_{k=1}^{n} f_k(z) \quad (z \in S).$$

If the sequence (F_n) tends uniformly in S to a function F, we say that the series $\sum_{n=1}^{\infty} f_n$ is **uniformly convergent**, or that it **sums to F uniformly in S**. Again, it is possible for a series to sum pointwise, but not uniformly: it follows immediately from Example 5.29 that the series $\sum_{n=1}^{\infty} z^n$ converges pointwise to $1/(1-z)$ in $N(0,1)$, but not uniformly.

The following result, known as the **Weierstrass**[4] **M-test**, is a useful tool for establishing the uniform convergence of a series.

---

[4] Karl Theodor Wilhelm Weierstrass, 1815–1897.

Theorem 5.31

For each $n \geq 1$, let f_n be a complex function with domain S, and suppose that there exist positive numbers M_n $(n \geq 1)$ such that $\|f_n\| \leq M_n$. If $\sum_{n=1}^{\infty} M_n$ is convergent, then $\sum_{n=1}^{\infty} f_n$ is uniformly convergent in S.

Proof

For each z in S and each $n \geq 1$

$$|f_n(z)| \leq \|f_n\| \leq M_n \,,$$

and so, by the Comparison Test, $\sum_{n=1}^{\infty} f_n(z)$, being absolutely convergent, is convergent. Denote its sum by $F(z)$ and its sum to N terms by $F_N(z)$. Let $\epsilon > 0$. Since $\sum_{n=1}^{\infty} M_n$ is convergent, there exists N such that, for all $n > N$,

$$\sum_{k=n+1}^{\infty} M_k < \epsilon/2 \,.$$

Hence, for all $m > n > N$ and all z in S,

$$\left| \sum_{k=n+1}^{m} f_k(z) \right| \leq \sum_{k=n+1}^{m} |f_k(z)| \leq \sum_{k=n+1}^{m} M_k < \epsilon/2 \,.$$

Letting m tend to ∞, we deduce that, for all z in S,

$$|F(z) - F_n(z)| \leq \epsilon/2 < \epsilon \,.$$

Hence $\|F - F_n\| < \epsilon$ for all $n > N$, and the proof is complete. $\qquad \square$

The following example should be compared with Example 5.29:

Example 5.32

Show that the geometric series $\sum_{n=0}^{\infty} z^n$ converges uniformly in any closed disc $\bar{N}(0, a)$, provided $0 \leq a < 1$.

Solution

For all z in $\bar{N}(0, a)$ and all $n \geq 1$,

$$|z^n| \leq a^n \,.$$

Since $\sum_{n=0}^{\infty} a^n$ is convergent, it follows by the M-test that $\sum_{n=0}^{\infty} z^n$ converges uniformly in $\bar{N}(0, a)$. $\qquad \square$

This is in fact a special case of a more general result concerning power series:

Theorem 5.33

Let $\sum_{n=0}^{\infty} c_n(z-a)^n$ be a power series with radius of convergence $R > 0$. Then, for all r in the interval $(0, R)$, the series is uniformly convergent in the closed disc $\bar{N}(a, r)$.

Proof

For all z in $\bar{N}(a, r)$, and for all n, we have that

$$|c_n(z-a)^n| \le |c_n r^n|.$$

Since $\sum_{n=0}^{\infty} |a_n r^n|$ is convergent by Theorem 4.14, the required conclusion follows immediately from the M-test. □

In Chapter 8 we shall have occasion to use the following result:

Theorem 5.34

Let γ be a piecewise smooth path, and suppose that (f_n) is a sequence of continuous functions, with common domain containing γ^*, such that $\sum_{n=1}^{\infty} f_n$ sums uniformly to a function F. Then

$$\int_\gamma \left[\sum_{n=1}^{\infty} f_n(z) \right] dz = \sum_{n=1}^{\infty} \left[\int_\gamma f_n(z)\, dz \right].$$

Proof

For each n, denote $\sum_{k=1}^{n} f_k$ by F_n, so that $F = \lim_{n \to \infty} F_n$. By Theorem 5.24,

$$\left| \int_\gamma F(z)\, dz - \sum_{k=1}^{n} \int_\gamma f_k(z)\, dz \right| = \left| \int_\gamma [F(z) - F_n(z)]\, dz \right|$$
$$\le L(\gamma^*)\|F - F_n\|,$$

where $L(\gamma^*)$ is the length of γ^*. By the assumption that $\sum_{n=1}^{\infty} f_n$ sums uniformly to F, this can be made less than any pre-assigned $\epsilon > 0$ by taking n sufficiently large. □

6
Cauchy's Theorem

6.1 Cauchy's Theorem: A First Approach

From Theorem 5.19, which can be seen as a complex version of the Fundamental Theorem of Calculus, we discern a strong tendency, when "reasonable" functions f and contours γ are involved, for $\int_\gamma f(z)\,dz$ to be zero. Corollary 5.20 mentions two special cases which we shall need to quote later, but many other familiar functions have the same property: for example, for a piecewise smooth contour γ,

$$\int_\gamma \sin z\,dz = \int_\gamma \cos z\,dz = \int_\gamma \exp z\,dz = 0\,.$$

(Simply observe that $\sin z = (-\cos)'(z)$, $\cos z = (\sin)'(z)$, $\exp z = (\exp)'(z)$.) The following general result occupies a central position in complex analysis:

Theorem 6.1 (Cauchy's Theorem)

Let γ^*, determined by a piecewise smooth function $\gamma : [a,b] \to \mathbb{C}$, be a contour, and let f be holomorphic in an open domain containing $\mathrm{I}(\gamma) \cup \gamma^*$. Then

$$\int_\gamma f(z)\,dz = 0\,.$$

How hard this is to prove depends on how general we want to be. In this chapter we first examine an approach that is adequate provided $\mathrm{I}(\gamma) \cup \gamma^*$ is either

(a) convex, or

(b) polygonal (whether convex or not).

In the next section we shall present a more difficult proof, which establishes the result for a general piecewise smooth contour.

We begin by showing that the theorem holds for a triangular contour:

Theorem 6.2

Let T be a triangular contour, and suppose that f is holomorphic in a domain containing $\mathrm{I}(T) \cup T$. Then $\int_T f(z)\, dz = 0$.

Proof

Let T be a triangle whose longest side is of length l, and suppose, for a contradiction, that

$$\left| \int_T f(z)\, dz \right| = h > 0.$$

We divide the triangle T into four equal subtriangles $\Delta_1, \Delta_2, \Delta_3, \Delta_4$, as shown:

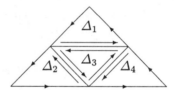

and (for $t = 1, 2, 3, 4$) let U_i be the boundary of Δ_i, oriented as shown. Observe now that

$$\int_T f(z)\, dz = \sum_{i=1}^{4} \int_{U_i} f(z)\, dz,$$

since on the right hand side each of the internal lines is traversed once in each direction, and so only the outer contour survives. Since

$$h = \left| \int_T f(z)\, dz \right| \leq \sum_{i=1}^{4} \left| \int_{U_i} f(z)\, dz \right|,$$

we must have

$$\left| \int_{U_i} f(z)\, dz \right| \geq \frac{h}{4}$$

for at least one of the triangular contours U_i. Choose one of these triangles, and rename it as T_1. Thus T_1, with longest side $l/2$, has the property that

$$\left| \int_{T_1} f(z)\, dz \right| \geq \frac{h}{4}\,.$$

We may repeat the process by subdividing T_1, choosing T_2, with longest side $l/4$, such that

$$\left| \int_{T_2} f(z)\, dz \right| \geq \frac{h}{16}\,;$$

then, continuing, we obtain, for each $n \geq 1$, a triangle T_n, with longest side $l/2^n$, such that

$$\left| \int_{T_n} f(z)\, dz \right| \geq \frac{h}{4^n}\,. \tag{6.1}$$

Much as in the proof of the Heine–Borel Theorem (Theorem 5.1), we can, for each n, select a point α_n within T_n and obtain a Cauchy sequence (α_n), with limit α lying inside *every* T_n.

Let $\epsilon > 0$ be given. From Theorem 4.11, there exists $\delta > 0$ such that

$$|f(z) - f(\alpha) - (z - \alpha)f'(\alpha)| < \epsilon|z - \alpha| \tag{6.2}$$

for all z in $N(\alpha, \delta)$. Choose n so that $T_n \subset N(\alpha, \delta)$.

By Corollary 5.20,

$$\int_{T_n} f(\alpha)\, dz = 0 \quad \text{and} \quad \int_{T_n} (z - \alpha)f'(\alpha)\, dz = 0\,.$$

Hence

$$\int_{T_n} f(z)\, dz = \int_{T_n} [f(z) - f(\alpha) - (z - \alpha)f'(\alpha)]\, dz\,.$$

Now, the perimeter of T_n is at most $3l/2^n$, and the maximum value of $|z - \alpha|$ for z and α in or on T_n is $l/2^n$. Hence, by (6.2) and Theorem 5.24,

$$\left| \int_{T_n} f(z)\, dz \right| \leq \epsilon\, \frac{3l}{2^n}\, \frac{l}{2^n} = \frac{3l^2\epsilon}{4^n}\,.$$

Comparing this with (6.1), we see that

$$h \leq 3l^2\epsilon\,.$$

Since ϵ can be chosen to be arbitrarily small, this gives a contradiction, and we are forced to conclude that

$$\int_T f(z)\, dz = 0\,.$$

$\square$

Corollary 6.3

Let γ be a piecewise smooth function determining a convex contour γ^*, and let f be holomorphic in an open domain containing $I(\gamma) \cup \gamma^*$. Then

$$\int_\gamma f(z)\,dz = 0\,.$$

Proof

From Theorems 6.2 and 5.25 we deduce that there exists a function F such that $F' = f$. Hence, by Theorem 5.19, $\int_\gamma f(z)\,dz = 0$.   □

Corollary 6.4

Let γ be a function determining a polygonal contour γ^*, and let f be holomorphic in an open domain containing $I(\gamma) \cup \gamma^*$. Then

$$\int_\gamma f(z)\,dz = 0\,.$$

Proof

The polygon can be divided into triangles $\Delta_1, \Delta_2, \ldots, \Delta_n$, with contours $T_1, T_2, \ldots, T_n$, respectively:

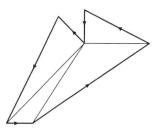

Then, by Theorem 6.2,

$$\int_\gamma f(z)\,dz = \sum_{i=1}^{n} \int_{T_i} f(z)\,dz = 0\,.$$

□

6.2 Cauchy's Theorem: A More General Version

Since we shall need to use Cauchy's Theorem and its consequences for contours that are neither convex nor polygonal, it becomes a duty on the author's part to present a proof of a more general case. Whether there is a corresponding duty on the reader's part is left to individual conscience! There is no doubt, however, that useful skills follow from the mastery of substantial proofs.

Proof* of Theorem 6.1

We begin by remarking that, by Corollary 6.4, Cauchy's Theorem is valid for any square or rectangular contour.

It will be convenient to use the notation $Q(\alpha, \epsilon)$ for the open square with sides parallel to the coordinate axes, centre at the point α and diagonal of length ϵ.

From the differentiability of f we deduce that, for every $\epsilon > 0$ and every α in $I(\gamma) \cup \gamma^*$, there exists $\delta_\alpha > 0$ with the property that

$$|f(z) - f(\alpha) - (z - \alpha)f'(\alpha)| < \epsilon|z - \alpha|$$

for all z in the open square $Q(\alpha, \delta_\alpha)$. From the open covering

$$\{Q(\alpha, \delta_\alpha) : \alpha \in I(\gamma) \cup \gamma^*\}$$

of the closed bounded set $I(\gamma) \cup \gamma^*$ we can, by the Heine–Borel Theorem (Theorem 5.1), select a finite subcovering

$$\{Q(\alpha_i, \delta_{\alpha_i}) : i = 1, 2, \ldots, N\};$$

then, simplifying the notation by writing δ_i rather than δ_{α_i}, we may assert that there exist points α_i $(i = 1, \ldots, N)$ with the property that

$$|f(z) - f(\alpha_i) - (z - \alpha_i)f'(\alpha_i)| < \epsilon|z - \alpha_i| \tag{6.3}$$

for all z in the open square $Q(\alpha_i, \delta_i)$.

Since the squares $Q_i = Q(\alpha_i, \delta_i)$ are open, a point on the boundary ∂Q_i of one square Q_i must lie properly inside another square Q_j. Hence there exists $\eta > 0$ with the property that, for all z in $I(\gamma) \cup \gamma^*$, there is a square Q_i for which $z \in Q_i$ and the distance $d(z, \partial Q_i) \geq \eta$.

Now suppose that $I(\gamma) \cup \gamma^*$ is covered by a square of side L, divided into smaller squares of side l by a grid of lines parallel to the coordinate axes, and choose $l < \eta/\sqrt{2}$.

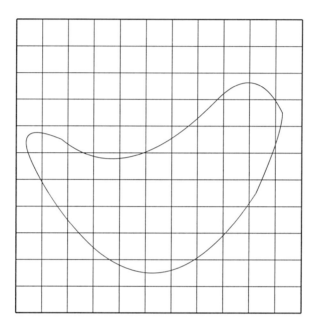

Lemma 6.5

Let Q be a square in the grid such that $Q \cap \big(I(\gamma) \cup \gamma^*\big) \neq \emptyset$. Then Q is contained in some Q_i.

Proof

Suppose, for a contradiction, that Q is not wholly contained in any one of $Q_1, Q_2, \ldots, Q_N$. Let $z \in Q$. By the covering property, z lies in at least one of the squares $Q_1, Q_2, \ldots, Q_N$. Choose one of those squares, and call it Q_i. By our assumption, there exists w in Q such that $w \notin Q_i$. By the covering property,

$w \in Q_j$ for some j.

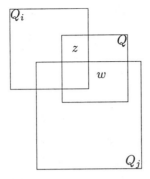

The line from z to w crosses the boundary of Q_i, and so, for every Q_i containing z,

$$d(z, \partial Q_i) < |z - w| \le l\sqrt{2} < \eta.$$

This is a contradiction, and so Q must be wholly contained in a single Q_i. □

Returning now to the main proof, we see that $I(\gamma)$ is thus divided into a set S of squares σ and a set T of incomplete squares τ, with boundaries $\partial\sigma$, $\partial\tau$, respectively. Each σ and each τ is contained in one of the squares $Q_1, Q_2, \ldots Q_N$ covering $I(\gamma) \cup \gamma^*$. We have lots of diagrams like

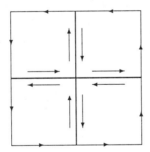

some involving incomplete squares, and because of all the internal cancellations we can assert that

$$\int_\gamma f(z)\, dz = \sum_{\sigma \in S} \int_{\partial\sigma} f(z)\, dz + \sum_{\tau \in T} \int_{\partial\tau} f(z)\, dz.$$

Hence, using Theorem 6.4, we deduce that

$$\int_\gamma f(z)\, dz = \sum_{\tau \in T} \int_{\partial\tau} f(z)\, dz. \tag{6.4}$$

By Theorem 5.5, our assumption about the function γ is sufficient to assure us that the curve γ^* is rectifiable, with total length Λ (say). Now consider a typical incomplete square τ, contained, by Lemma 6.5, in an open square Q_i,

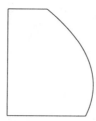

and suppose that the length of the piece of contour forming part of the boundary of τ is λ_τ. By Lemma 6.5, τ is contained in a square $Q_i = Q(\alpha_i, \delta_i)$ with the property that

$$|f(z) - f(\alpha_i) - (z - \alpha_i)| < \epsilon|z - \alpha_i|$$

for every z in Q_i, and so certainly for every z in τ. By a now familiar argument, we deduce that

$$\left| \int_{\partial\tau} f(z)\, dz \right| \leq \int_{\partial\tau} \epsilon|z - \alpha_i|\, dz\,.$$

The total length of the boundary of τ is certainly not greater than $4l + \lambda_\tau$, and $|z - \alpha_i|$ cannot exceed $l\sqrt{2}$. So, by Theorem 5.24,

$$\left| \int_{\partial\tau} f(z)\, dz \right| \leq (4l + \lambda_\tau)\epsilon l\sqrt{2} < 4(l^2 + l\lambda_\tau)\epsilon\sqrt{2}\,.$$

Summing over all τ gives

$$\left| \int_\gamma f(z)\, dz \right| \leq 4(A + l\Lambda)\epsilon\sqrt{2}\,,$$

where A is the area of the outer, bounding square, and Λ is the length of the contour γ^*. Since the expression on the right can be made arbitrarily small, we are forced to conclude that

$$\int_\gamma f(z)\, dz = 0\,.$$

$\square$

6.3 Deformation

As we shall see, the consequences of Cauchy's Theorem are many and important. We end this chapter with some of the most obvious corollaries to the result.

Theorem 6.6

Let $\gamma_1, \gamma_2 : [a, b] \to \mathbb{C}$ be piecewise smooth curves such that

$$\gamma_1(a) = \gamma_2(a), \quad \gamma_1(b) = \gamma_2(b), \quad \gamma_1(t_1) \neq \gamma_2(t_2) \quad (t_1, t_2 \in (a, b)).$$

If f is holomorphic throughout an open set containing γ_1^*, γ_2^* and the region between, then

$$\int_{\gamma_1} f(z)\,dz = \int_{\gamma_2} f(z)\,dz.$$

Proof

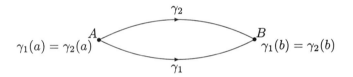

$\gamma_1(a) = \gamma_2(a)$ and $\gamma_1(b) = \gamma_2(b)$

Let σ^* be the simple closed curve travelling from A to B via γ_1 and from B to A via $-\gamma_2$. Since f is holomorphic in an open domain containing $\mathrm{I}(\sigma) \cup \sigma^*$, we have that

$$0 = \int_{\sigma} f(z)\,dz = \int_{\gamma_1} f(z)\,dz - \int_{\gamma_2} f(z)\,dz.$$

$\square$

Finally, we have

Theorem 6.7 (The Deformation Theorem)

Let γ_1, γ_2 be contours, with γ_2 lying wholly inside γ_1, and suppose that f is holomorphic in a domain containing the region between γ_1 and γ_2. Then

$$\int_{\gamma_1} f(z)\,dz = \int_{\gamma_2} f(z)\,dz.$$

Proof

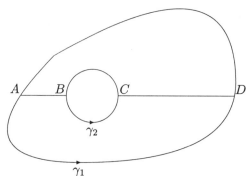

Join the two contours by lines AB and CD as shown. Denote the (lower) section of γ_1, from A to D, by γ_{1l} and the (upper) section, from D to A, by γ_{1u}. Similarly, denote the lower section of γ_2 from B to C, by γ_{2l}, and the upper section, from C to B, by γ_{2u}.

Form a contour σ_1 by traversing from A to B, then from B to C by $-\gamma_{2u}$, then from C to D, and finally from D back to A by γ_{1u}. By assumption, the function f is holomorphic inside and on σ_1, and so $\int_{\sigma_1} f(z)\, dz = 0$. That is,

$$\int_{AB} f(z)\, dz - \int_{\gamma_{2u}} f(z)\, dz + \int_{CD} f(z)\, dz + \int_{\gamma_{1u}} f(z)\, dz = 0\,. \qquad (6.5)$$

Similarly, form a contour σ_2 by traversing from A to D by γ_{1l}, then from D to C, then from C to B by $-\gamma_{2l}$, and finally from B back to A. Thus

$$\int_{\gamma_{1l}} f(z)\, dz - \int_{CD} f(z)\, dz - \int_{\gamma_{2l}} f(z)\, dz - \int_{AB} f(z)\, dz = 0\,. \qquad (6.6)$$

Adding (6.5) and (6.6) gives

$$\left(\int_{\gamma_{1u}} f(z)\, dz + \int_{\gamma_{1l}} f(z)\, dz \right) - \left(\int_{\gamma_{2u}} f(z)\, dz + \int_{\gamma_{2l}} f(z)\, dz \right) = 0\,,$$

and so

$$\int_{\gamma_1} f(z)\, dz = \int_{\gamma_2} f(z)\, dz\,,$$

as required.

$\square$

EXERCISES

6.1. Let γ be a contour such that $0 \in I(\gamma)$. Show that

$$\int_\gamma z^n \, dz = \begin{cases} 2\pi i & \text{if } n = -1 \\ 0 & \text{otherwise.} \end{cases}$$

By taking γ^* as the ellipse

$$\left\{ (x, y) : \frac{x^2}{a^2} + \frac{y^2}{b^2} = 1 \right\},$$

show that

$$\int_0^{2\pi} \frac{dt}{a^2 \cos^2 t + b^2 \sin^2 t} = \frac{2\pi}{ab}.$$

6.2. Look again at Exercise 5.11, and obtain the improved bound

$$\left| \int_\gamma \sin(z^2) \, dz \right| \le 2a \cosh(4a^2).$$

6.3. By applying Cauchy's Theorem to e^z and integrating round a circular contour, show that

$$\int_0^{2\pi} e^{r \cos \theta} \cos(r \sin \theta + \theta) \, d\theta = 0.$$

Some Consequences of Cauchy's Theorem

7.1 Cauchy's Integral Formula

We have already observed in Theorem 5.13 that if σ is a circle with centre 0 then

$$\int_\sigma \frac{1}{z}\, dz = 2\pi i\,.$$

More generally, if $\kappa(a, r)$ is a circle with centre a, then

$$\int_{\kappa(a,r)} \frac{1}{z - a}\, dz = 2\pi i\,; \tag{7.1}$$

for, with $z = a + re^{i\theta}$,

$$\int_{\kappa(a,r)} \frac{1}{z - a}\, dz = \int_0^{2\pi} \frac{ire^{i\theta}\, d\theta}{re^{i\theta}} = 2\pi i\,.$$

This observation will play a part in the proof of the main result of this section, which shows that *the value of a holomorphic function inside a contour is determined by its values on the contour.* (Recall that by a **contour** we shall always mean a *closed, simple, piecewise smooth curve*, and that, unless we specify otherwise, it is traversed in the positive direction.) That is an extraordinary result, and reveals a fundamental difference between complex analysis and real analysis. Even for an analytic real function f (infinitely differentiable and with a Taylor series expansion) we can make no deduction at all about the values of the function in (a, b) from its values at a and at b.

Theorem 7.1 (Cauchy's Integral Formula)

Let γ be a contour and let f be holomorphic in an open domain containing $I(\gamma) \cup \gamma^*$. Then for every point a in $I(\gamma)$,

$$f(a) = \frac{1}{2\pi i} \int_\gamma \frac{f(z)}{z-a}\, dz\,.$$

Proof

Let $a \in I(\gamma)$. By the differentiability of f at a we know (see Theorem 4.11) that

$$f(z) = f(a) + (z-a)f'(a) + v(z,a)(z-a)\,, \qquad (7.2)$$

where $v(z,a)$ tends to 0 as $z \to a$. That is, for all $\epsilon > 0$ there exists $\delta > 0$ such that $|v(z,a)| < \epsilon$ for all z in $N(a,\delta)$.

Let $\kappa = \kappa(a,r)$, the circle with centre a and radius r, where r is chosen so that

(i) the disc $N(a,r)$ lies wholly inside γ; and

(ii) $r < \delta$.

Since $f(z)/(z-a)$ is holomorphic in the region between σ and γ we deduce by the Deformation Theorem (Theorem 6.7) that

$$\int_\gamma \frac{f(z)}{z-a}\, dz = \int_\kappa \frac{f(z)}{z-a}\, dz$$

$$= f(a) \int_\kappa \frac{1}{z-a}\, dz + f'(a) \int_\kappa 1\, dz + \int_\kappa v(z,a)\, dz \quad \text{(by (7.2))}$$

$$= 2\pi i\, f(a) + \int_\kappa v(z,a)\, dz \quad \text{(by (7.1) and Corollary 5.19)}\,.$$

Thus

$$\left| \int_\gamma \frac{f(z)}{z-a}\, dz - 2\pi i\, f(a) \right| = \left| \int_\kappa v(z,a)\, dz \right| < 2\pi r \epsilon \quad \text{(by Theorem 5.24)}\,.$$

Since this holds for *every* positive ϵ, we deduce that

$$f(a) = \frac{1}{2\pi i} \int_\gamma \frac{f(z)}{z-a}\, dz\,.$$

$\square$

Remark 7.2

Dividing $f(z)$ by $z - a$ introduces a singularity (unless $f(a) = 0$). Cauchy's Integral Formula is the first indication that integration round a contour depends crucially on the singularities of the integrand within the contour.

Example 7.3

Evaluate

$$\int_{\kappa(0,2)} \frac{\sin z}{z^2 + 1} \, dz \, .$$

Solution

Since

$$\frac{1}{z^2 + 1} = \frac{1}{2i} \left(\frac{1}{z - i} - \frac{1}{z + i} \right) ,$$

we may deduce that

$$\int_{\kappa(0,2)} \frac{\sin z}{z^2 + 1} \, dz = \frac{1}{2i} \int_{\kappa(0,2)} \frac{\sin z}{z - i} \, dz - \frac{1}{2i} \int_{\kappa(0,2)} \frac{\sin z}{z + i}$$

$$= \pi(\sin i - \sin(-i)) = 2\pi \sin i = \frac{\pi}{i} \left(e^{-1} - e \right) = \pi i(e - e^{-1}) .$$

□

If we could be sure that the procedure of differentiating under the integral sign was valid, we could deduce from Cauchy's Integral Formula that

$$f'(a) = \frac{1}{2\pi i} \int_\gamma \frac{d}{da} \left(\frac{f(z)}{z - a} \right) \, dz = \frac{1}{2\pi i} \int_\gamma \frac{f(z)}{(z - a)^2} \, dz \, .$$

In fact this is true, though what we have just written does not even approximate to a proof.

Theorem 7.4

Let γ be a contour, let f be holomorphic in a domain containing $I(\gamma) \cup \gamma^*$, and let $a \in I(\gamma)$. Then

$$f'(a) = \frac{1}{2\pi i} \int_\gamma \frac{f(z)}{(z - a)^2} \, dz \, .$$

Proof

Since $a \in I(\gamma)$, there exists $\delta > 0$ such that

$$|z - a| > 2\delta \tag{7.3}$$

for all z on the contour γ^*. If $0 < |h| < \delta$, then $a + h \in I(\gamma)$ and, for all z on γ^*,

$$|z - a - h| \geq |z - a| - |h| > \delta.\qquad(7.4)$$

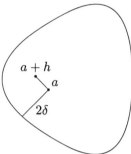

Since f is continuous on the closed, bounded set γ^*, it follows by Theorem 5.3 that the set

$$\{|f(z)| : z \in \gamma^*\}$$

is bounded, with supremum M (say).

By Cauchy's Integral Formula,

$$\frac{f(a + h) - f(a)}{h} = \frac{1}{2\pi i h} \int_\gamma f(z) \left(\frac{1}{z - a - h} - \frac{1}{z - a}\right) dz$$

$$= \frac{1}{2\pi i} \int_\gamma \frac{f(z)}{(z - a - h)(z - a)} \, dz.$$

Hence

$$\left| \frac{f(a + h) - f(a)}{h} - \frac{1}{2\pi i} \int_\gamma \frac{f(z)}{(z - a)^2} \, dz \right|$$

$$= \left| \frac{1}{2\pi i} \int_\gamma f(z) \left[\frac{1}{(z - a - h)(z - a)} - \frac{1}{(z - a)^2} \right] dz \right|$$

$$= \frac{|h|}{2\pi} \left| \int_\gamma \frac{f(z)}{(z - a)^2(z - a - h)} \, dz \right|.$$

By Theorem 5.24 and Equations (7.3) and (7.4) we now conclude that

$$\left| \frac{f(a + h) - f(a)}{h} - \frac{1}{2\pi i} \int_\gamma \frac{f(z)}{(z - a)^2} \, dz \right| < \frac{ML|h|}{8\pi\delta^3},$$

where L is the length of the contour γ. Thus

$$\lim_{h \to 0} \left| \frac{f(a + h) - f(a)}{h} - \frac{1}{2\pi i} \int_\gamma \frac{f(z)}{(z - a)^2} \, dz \right| = 0,$$

and so

$$f'(a) = \frac{1}{2\pi i} \int_\gamma \frac{f(z)}{(z - a)^2} \, dz,$$

as required.   □

More generally, we have the following theorem:

Theorem 7.5

Let γ be a contour, let f be holomorphic in a domain containing $\mathrm{I}(\gamma) \cup \gamma^*$, and let $a \in \mathrm{I}(\gamma)$. Then f has an nth derivative $f^{(n)}$ for all $n \geq 1$, and

$$f^{(n)}(a) = \frac{n!}{2\pi i} \int_\gamma \frac{f(z)}{(z-a)^{n+1}} \, dz \, .$$

Proof*

The proof is by induction on n, and we have already proved the result for $n = 1$. We suppose that the result holds for $n = k - 1$ and consider the expression $[f^{(k-1)}(a+h) - f^{(k-1)}(a)]/h$. As in the proof of Theorem 7.4, we can find $\delta > 0$ and can choose h such that $|z - a| > 2\delta$, $|z - a - h| > \delta$. By the induction hypothesis,

$$\frac{f^{(k-1)}(a+h) - f^{(k-1)}(a)}{h} = \frac{(k-1)!}{2\pi i h} \int_\gamma f(z) \left[\frac{1}{(z-a-h)^k} - \frac{1}{(z-a)^k} \right] dz \, .$$

To prove the theorem, we need to show that $E(h)$, defined by

$$E(h) = \frac{f^{(k-1)}(a+h) - f^{(k-1)}(a)}{h} - \frac{k!}{2\pi i} \int_\gamma \frac{f(z)}{(z-a)^{k+1}} \, dz \, ,$$

tends to 0 as $h \to 0$. Now,

$$E(h) = \frac{(k-1)!}{2\pi i h} \int_\gamma f(z) \left[\frac{1}{(z-a-h)^k} - \frac{1}{(z-a)^k} - \frac{kh}{(z-a)^{k+1}} \right] dz \, ,$$

and

$$D = \frac{1}{(z-a-h)^k} - \frac{1}{(z-a)^k} - \frac{kh}{(z-a)^{k+1}} = g(a+h) - g(a) - hg'(a) \, ,$$

where $g(w) = 1/(z-w)^k$. Hence, by (4.5), we know that $D = hv(a,h)$, where $v(a,h) \to 0$ as $h \to 0$.

There exists M such that $|f(z)| \leq M$ in $\mathrm{I}(\gamma) \cup \gamma^*$, and the contour γ^* has length L (say). Hence, by Theorem 5.24,

$$|E(h)| \leq \frac{(k-1)! M L |v(a,h)|}{2\pi} \, ,$$

which tends to 0 as $h \to 0$. $\qquad\square$

Remark 7.6

It is worth drawing attention to the fact, on the face of it rather surprising, that a differentiable function f necessarily has higher derivatives of every order. This is in complete contrast to the situation in real analysis, where, for example, the function f given by

$$f(x) = \begin{cases} x^2 \sin(1/x) & \text{if } x \neq 0 \\ 0 & \text{if } x = 0 \end{cases}$$

is differentiable at 0, but f' is not even continuous.

The statement of Theorem 7.5 tends to suggest that one wants to use the integral to obtain the derivative. Frequently, however, it is appropriate to turn the formula round, and to use the equality

$$\int_\gamma \frac{f(z)}{(z-a)^n} = \frac{2\pi i}{(n-1)!} f^{(n-1)}(a) \tag{7.5}$$

to compute the integral.

Example 7.7

Evaluate

$$\int_{\kappa(0,1)} \frac{e^{\sin z}}{z^3} \, dz \, .$$

Solution

By Theorem 7.5, the value of the integral is $(1/2!)2\pi i f''(0)$, where $f(z) = e^{\sin z}$. Now, $f'(z) = e^{\sin z} \cos z$, and $f''(z) = e^{\sin z}(\cos^2 z - \sin z)$. Thus $f''(0) = 1$, and so

$$\int_{\kappa(0,1)} \frac{e^{\sin z}}{z^3} \, dz = \pi i \, .$$

$\square$

We finish this section by showing that Cauchy's Theorem has a converse:

Theorem 7.8 (Morera's Theorem)

Let f be continuous on an open set D. If $\int_\gamma f(z) \, dz = 0$ for every contour contained in D, then f is holomorphic in D.

Proof

Let $a \in D$, and let r be such that $N(a, r) \subseteq D$. Within this convex open set every contour γ, in particular every triangular contour γ, is such that $\int_{\gamma} f(z)\, dz = 0$. Hence, by Theorem 5.19, there exists a function F, holomorphic in $N(a, r)$, such that $F'(z) = f(z)$ for all z in $N(a, r)$. By the remark following Theorem 7.5, F has derivatives of all orders within $N(a, r)$, and so certainly $f'(a)$ exists. Since a was chosen arbitrarily, it follows that f is holomorphic in D. $\qquad\qquad\qquad\square$

EXERCISES

7.1. Evaluate the following integrals:

a) $\displaystyle \int_{\kappa(0,1)} \frac{e^{kz}}{z^{n+1}}\, dz$;

b) $\displaystyle \int_{\kappa(0,2)} \frac{z^3}{z^2 - 2z + 2}\, dz$;

c) $\displaystyle \int_{\kappa(0,2)} \frac{e^z}{\pi i - 2z}\, dz$.

7.2. Evaluate
$$\int_{\kappa(0,2)} \frac{z^m}{(1-z)^n}\, dz \quad (m, n \in \mathbb{N}).$$

7.3. Suppose that the function f is holomorphic in $N(a, R)$. Show that, if $0 < r < R$,
$$f'(a) = \frac{1}{\pi r} \int_0^{2\pi} F(\theta) e^{-i\theta}\, d\theta$$
where $F(\theta)$ is the real part of $f(a + re^{i\theta})$.

7.4. Suppose that the function f is holomorphic in $N(0, R')$, and let a be such that $|a| = r < R < R'$.

a) Show that
$$f(a) = \frac{1}{2\pi i} \int_{\kappa(0,R)} \frac{R^2 - a\bar{a}}{(z - a)(R^2 - z\bar{a})}\, f(z)\, dz.$$

b) Deduce **Poisson's**[1] formula: if $0 < r < R$, then
$$f(re^{i\theta}) = \frac{1}{2\pi} \int_0^{2\pi} \frac{R^2 - r^2}{R^2 - 2Rr\cos(\theta - \phi) + r^2}\, f(Re^{i\phi})\, d\phi.$$

---

[1] Siméon Denis Poisson, 1781–1840.

7.5. Evaluate

$$\int_{\kappa(0,4)} \frac{\sin^2 z \, dz}{[z - (\pi/6)]^2 [z + (\pi/6)]} \, .$$

7.6. Prove the following result:

Let γ be a contour, and let f be continuous on γ^*. Then g, defined by

$$g(z) = \int_\gamma \frac{f(w)}{w - z} \, dz \, ,$$

is holomorphic in $\mathbb{C} \setminus \gamma^*$.

7.2 The Fundamental Theorem of Algebra

Recall now that a function f which is holomorphic throughout $\mathbb{C}$ will be called an **entire** function. We have already encountered several such functions: every polynomial function is an entire function, and so are exp, sin and cos. When regarded as real functions, sin and cos are also bounded, but the boundedness property fails when we consider the whole complex plane: both $|\cos(iy)| = \cosh y$ and $|\sin(iy)| = |i \sinh y|$ (where y is real) tend to infinity as $y \to \infty$. It is thus natural to ask whether there exist any bounded entire functions. Liouville's Theorem[2] says in essence that there are none:

Theorem 7.9 (Liouville's Theorem)

Let f be a bounded entire function. Then f is constant.

Proof

Suppose that $|f(z)| \leq M$ for all z in $\mathbb{C}$. Let $a \in \mathbb{C}$ and let γ_R be the circular contour $|z - a| = R$. Then, by Theorems 7.4 and 5.24,

$$|f'(a)| = \left| \frac{1}{2\pi i} \int_{\gamma_R} \frac{f(z)}{(z-a)^2} \, dz \right| \leq \frac{1}{2\pi} \cdot \frac{M}{R^2} \cdot 2\pi R = \frac{M}{R} \, .$$

This holds for all values of R, and so $f'(a) = 0$. Since $f'(a) = 0$ for all a in $\mathbb{C}$, it follows from Theorem 4.9 that f is a constant function. $\square$

---

[2] Joseph Liouville, 1809–1882.

The rather grandly named Fundamental Theorem of Algebra has already been mentioned (see Section 2.1) as one of the justifications for studying complex numbers. It was proved first by Gauss[3], and is a fine example of a deep and difficult theorem that yields easily once we have developed suitable tools. For a history of the theorem, see [11].

Theorem 7.10

Let $p(z)$ be a polynomial of degree $n \geq 1$, with coefficients in $\mathbb{C}$. Then there exists a in $\mathbb{C}$ such that $p(a) = 0$.

Proof

Suppose, for a contradiction, that $p(z) \neq 0$ for all z in $\mathbb{C}$. Then both $p(z)$ and $1/p(z)$ are entire functions. Certainly $|p(z)| \to \infty$ as $|z| \to \infty$ (see Exercise 3.6), and so there exists $R > 0$ such that $|1/p(z)| \leq 1$ whenever $|z| > R$. By Theorem 5.3, the function $1/p(z)$ is also bounded on the closed bounded set $\{z : |z| \leq R\}$. Thus, by Theorem 7.9, $1/p(z)$, being a bounded entire function, must be constant. From this contradiction we deduce, as required, that the polynomial equation $p(z) = 0$ must have at least one root. $\square$

It is now straightforward to prove

Theorem 7.11 (The Fundamental Theorem of Algebra)

Let $p(z)$ be a polynomial of degree n, with coefficients in $\mathbb{C}$. Then there exist complex numbers $\beta, \alpha_1, \alpha_2, \ldots, \alpha_n$ such that

$$p(z) = \beta(z - \alpha_1)(z - \alpha_2)\ldots(z - \alpha_n).$$

Proof

The proof is by induction on n, it being clear that the result is valid for $n = 1$. Suppose that the result holds for all polynomials of degree $n - 1$, and let $p(z)$ have degree n. By Theorem 7.10 there exists α_1 in $\mathbb{C}$ such that $p(\alpha_1) = 0$. Hence $p(z) = (z - \alpha_1)q(z)$, where $q(z)$ is of degree $n - 1$. By the induction hypothesis there exist $\beta, \alpha_2, \ldots, \alpha_n$ in $\mathbb{C}$ such that

$$q(z) = \beta(z - \alpha_2)\ldots(z - \alpha_n).$$

---

[3] Johann Carl Friedrich Gauss, 1777–1855.

Hence

$$p(z) = \beta(z - \alpha_1)(z - \alpha_2)\ldots(z - \alpha_n)\,,$$

as required. □

EXERCISES

7.7. Let $p(x) = a_0 + a_1 x + \cdots + a_n x^n$ be a polynomial of degree n with real coefficients. Show that $p(x)$ factorises into linear and quadratic factors. That is, show that, for some $k, l \geq 0$ such that $k + 2l = n$, there exist real numbers $\alpha_1, \alpha_2, \ldots, \alpha_k,\ \beta_1, \beta_2, \ldots, \beta_l,\ \gamma_1, \gamma_2, \ldots, \gamma_l$ such that

$$p(x) = a_n(x - \alpha_1)\ldots(x - \alpha_k)(x^2 + \beta_1 x + \gamma_1)\ldots(x^2 + \beta_l x + \gamma_l)\,.$$

Deduce that a polynomial of odd degree must have at least one real root.

7.8 Find the real factors of

$$x^6 + 1, \quad x^4 - 3x^3 + 4x^2 - 6x + 4, \quad x^4 + 3x^3 - 3x^2 - 7x + 6\,.$$

7.3 Logarithms

We have already encountered the logarithm function in Section 3.5, where we discussed the problem of finding w such that $e^w = z$. A related problem is that of finding, for a contour γ, such that $0 \notin I(\gamma) \cup \gamma^*$, a function $\log_\gamma$, holomorphic in $I(\gamma)$, such that $\exp[\log_\gamma z] = z$.

This follows from a theorem concerning what, following the terminology of real analysis, we might call the **Indefinite Integral Theorem**:

Theorem 7.12

Let γ be a contour and let f be holomorphic in an open domain containing $I(\gamma) \cup \gamma^*$. Then there exists a holomorphic function F such that $F'(z) = f(z)$ for all z in $I(\gamma)$.

Proof

Choose and fix a point a in $I(\gamma)$, and, for each z in $I(\gamma)$, let δ_z be a smooth curve from a to z, lying wholly within $I(\gamma)$. Let

$$F(z) = \int_{\delta_z} f(w)\,dw\,;$$

by the deformation theorem (Theorem 6.7), this is independent of the precise curve we choose from a to z. Choose h so that $N(z, |h|)$ lies wholly in $I(\gamma)$. Then certainly the line segment $[z, z + h]$ lies within $I(\gamma)$.

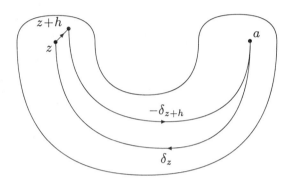

We can certainly arrange for the closed path $\left(\delta_z, [z, z + h], -\delta_{z+h}\right)$ to be simple (that is, without crossings), and so, by Cauchy's Theorem,

$$F(z) + \int_{[z,z+h]} f(w)\,dw - F(z + h) = 0\,.$$

Hence

$$F(z + h) - F(z) = \int_{[z,z+h]} f(w)\,dw\,.$$

Now,

$$\int_{[z,z+h]} f(z)\,dw = f(z) \int_{[z,z+h]} 1\,dw = hf(z)\,,$$

and so, by Theorem 5.24,

$$\left| \frac{F(z+h) - F(z)}{h} - f(z) \right| = \frac{1}{|h|} \left| \int_{[z,z+h]} \big(f(w) - f(z)\big)\,dw \right|$$

$$\leq \frac{1}{|h|} \left[|h| \sup_{w \in [z,z+h]} |f(w) - f(z)| \right]\,.$$

Since, by the continuity of f, this tends to 0 as $h \to 0$, we deduce that F is differentiable at z, with derivative $f(z)$. $\square$

Example 7.13

Show that there exists F such that $F'(z) = e^{z^2}$ within the neighbourhood $N(0, R)$.

Solution

By the theorem just proved, the required function is given by

$$F(z) = \int_\gamma e^{w^2}\, dw\,,$$

where γ is any smooth path from 0 to z. The fact that we cannot "do" the integral by antidifferentiation has nothing to do with the existence of the function. □

Theorem 7.14

Let D be an open disc not containing 0. Then there exists a function F, holomorphic in D, such that

$$e^{F(z)} = z \quad (z \in D)\,.$$

Proof

Let a be an arbitrary fixed point in D and, for each z in D, let δ_z be a smooth path in D from a to z. The function $1/z$ is holomorphic in $\mathbb{C} \setminus \{0\}$ and so, by Theorem 7.12, the function G given by

$$G(z) = \int_{\delta_z} \frac{1}{w}\, dw$$

has the property that $G'(z) = 1/z$. Let $H(z) = ze^{-G(z)}$. Then, for all z in D,

$$H'(z) = e^{-G(z)} - ze^{-G(z)}G'(z) = e^{-G(z)} - e^{-G(z)} = 0\,,$$

and so, by Theorem 4.9, $e^{G(z)} = Cz$, for some constant C. Let $F(z) = G(z) - \log C$; then $e^{F(z)} = z$, as required. □

Remark 7.15

It is reasonable to denote the holomorphic function F as $\log_D z$. Like the logarithm of a number, it is not quite unique. If $e^{F_1(z)} = e^{F_2(z)} (= z)$ for all z in D, then differentiation gives

$$zF_1'(z) = e^{F_1(z)}F_1'(z) = e^{F_2(z)}F_2'(z) = zF_2'(z)\,.$$

Thus $F_1 - F_2$ has zero derivative and so is a constant K. Since $e^K = 1$ we must have $K = 2n\pi i$ for some integer n:

$$F_1(z) - F_2(z) = 2n\pi i \quad (z \in D).$$

7.4 Taylor Series

In real analysis there are distinctions between functions that are **differentiable**, **infinitely differentiable** (having derivatives of all orders), and **analytic** (having a Taylor[4] series expansion). We have already seen that a holomorphic function is infinitely differentiable. In fact it also has a Taylor series expansion. Precisely, we have the following theorem:

Theorem 7.16

Let $c \in \mathbb{C}$ and suppose that the function f is holomorphic in some neighbourhood $N(c, R)$ of c. Then, within $N(c, R)$,

$$f(z) = \sum_{n=0}^{\infty} a_n (z - c)^n,$$

where, for $n = 0, 1, 2, \ldots$,

$$a_n = \frac{f^{(n)}(c)}{n!}.$$

Proof

It is helpful first to record the sum of the following finite geometric series:

$$\frac{1}{z - c} + \frac{h}{(z - c)^2} + \cdots + \frac{h^n}{(z - c)^{n+1}} = \frac{1}{z - c - h} - \frac{h^{n+1}}{(z - c)^{n+1}(z - c - h)}. \quad (7.6)$$

Let $0 < R_1 < R_2 < R$. Then f is holomorphic throughout the closed disc $\bar{N}(c, R_2)$. Let C be the circle $\kappa(c, R_2)$, and let $c + h \in \bar{N}(c, R_1)$. Then, by Theorems 7.1 and 7.5 and Equation (7.6),

$$f(c + h) = \frac{1}{2\pi i} \int_C \frac{f(z)}{z - c - h}$$

$$= \frac{1}{2\pi i} \left[\int_C \frac{f(z)}{z - c} \, dz + h \int_C \frac{f(z)}{(z - c)^2} \, dz + \cdots + h^n \int_C \frac{f(z)}{(z - c)^{n+1}} \, dz \right.$$

---

[4] Brook Taylor, 1685–1731.

$$+ h^{n+1} \int_C \frac{f(z)}{(z-c)^{n+1}(z-c-h)} \, dz \Big]$$

$$= f(c) + hf'(c) + \cdots + \frac{h^n}{n!} f^{(n)}(c) + E_n \, ,$$

where

$$E_n = \frac{h^{n+1}}{2\pi i} \int_C \frac{f(z)}{(z-c)^{n+1}(z-c-h)} \, dz \, .$$

We complete the proof by showing that $E_n \to 0$ as $n \to \infty$. First, by Theorem 5.3, there exists $M > 0$ such that $|f(z)| \le M$ for all z on the circle C. For all z on C,

$$|z - c - h| \ge |z - c| - |h| \ge R_2 - R_1 \, ,$$

since $|z - c| = R_2$ and $|h| \le R_1$. Hence, by Theorem 5.24,

$$|E_n| \le \frac{|h|^{n+1}}{2\pi} \cdot 2\pi R_2 \cdot \frac{M}{R_2^{n+1}(R_2 - R_1)} = \frac{M|h|}{R_2 - R_1} \left(\frac{|h|}{R_2} \right)^n \, .$$

Since $|h|/R_2 < 1$, we deduce that $E_n \to 0$ as $n \to \infty$. Thus

$$f(c + h) = \sum_{n=0}^{\infty} \frac{h^n}{n!} f^{(n)}(c) \, ,$$

and substituting $z = c + h$ gives

$$f(z) = \sum_{n=0}^{\infty} a_n (z - c)^n \, ,$$

where

$$a_n = \frac{1}{n!} f^{(n)}(c) = \frac{1}{2\pi i} \int_C \frac{f(z)}{(z-c)^{n+1}} \, dz \, . \qquad (7.7)$$

$\square$

Remark 7.17

The Taylor series of a function f is unique. If $f(z) = \sum_{n=0}^{\infty} a_n (z - c)^n$ then, by Theorem 4.19, $f^{(n)}(z) = n! a_n +$ positive powers of $z - c$, and so $a_n = f^{(n)}(c)/n!$. Thus if we find, by whatever method, a power series for a function, the series we find must be the Taylor series.

Example 7.18

Show that, for all real a,

$$(1 + z)^a = 1 + \sum_{n=1}^{\infty} \frac{a(a-1)\ldots(a-n+1)}{n!} z^n \quad (|z| < 1) \, .$$

Solution

The function is holomorphic in the open set $\mathbb{C} \setminus \{-1\}$, so certainly in the neighbourhood $N(0, 1)$. Moreover, the principal argument of $1 + z$ lies safely in the interval $(-\pi/2, \pi/2)$, and so there is no ambiguity in the meaning of $(1 + z)^a = e^{a \log(1+z)}$. A routine calculation gives

$$f^{(n)}(z) = a(a - 1) \dots (a - n + 1)(1 + z)^{a-n} \quad (n = 1, 2, \dots)$$

and so

$$\frac{f^{(n)}(0)}{n!} = \frac{a(a - 1) \dots (a - n + 1)}{n!}.$$

Hence

$$(1 + z)^a = 1 + \sum_{n=1}^{\infty} \frac{a(a - 1) \dots (a - n + 1)}{n!} z^n \quad (|z| < 1)$$

as required. $\square$

We sometimes want to say that the Taylor series

$$f(z) = \sum_{n=0}^{\infty} a_n (z - c)^n$$

is the **Taylor series of f at c**, or that the series is **centred** on c, or that c is the **centre** of the series. To qualify Remark 7.17 above, the Taylor series of a function is unique *once we choose the centre*, but a function has many different Taylor series, with different centres. For example, the function $1/(1 + z)^2$ is holomorphic in $\mathbb{C} \setminus \{-1\}$, and its Taylor series, centred on 0, is

$$1 - 2z + 3z^2 - 4z^3 + \cdots = \sum_{n=0}^{\infty} (-1)^n (n + 1) z^n \quad (|z| < 1).$$

If we choose an arbitrary complex number $c \neq -1$ as centre, we find that

$$\frac{1}{(1 + z)^2} = \frac{1}{[(c + 1) + (z - c)]^2} = \frac{1}{(c + 1)^2 [1 + ((z - c)/(c + 1))]^2}$$

$$= \sum_{n=0}^{\infty} \frac{(-1)^n (n + 1)}{(c + 1)^{n+2}} (z - c)^n \quad (|z - c| < |c + 1|).$$

Example 7.19

Show that

$$\log(1 + z) = z - \frac{z^2}{2} + \frac{z^3}{3} - \cdots \quad (|z| < 1).$$

Solution

Within the neighbourhood $N(0,1)$, $1+z$ stays clear of the cut along the negative x-axis.

$$
\begin{aligned}
f(z) &= \log(1+z) & f(0) &= 0 \\
f'(z) &= \frac{1}{1+z} & f'(0) &= 1 \\
f''(z) &= -\frac{1}{(1+z)^2} & \frac{f''(0)}{2!} &= \frac{1}{2} \\
f'''(z) &= \frac{2}{(1+z)^3} & \frac{f'''(0)}{3!} &= \frac{1}{3} \\
& \quad \cdots \\
f^{(n)}(z) &= (-1)^{n-1}\frac{(n-1)!}{(1+z)^n} & \frac{f^{(n)}(0)}{n!} &= \frac{(-1)^{n-1}}{n},
\end{aligned}
$$

and so

$$
\log(1+z) = \sum_{n=1}^{\infty} (-1)^{n-1} \frac{z^n}{n} \quad (|z| < 1).
$$

$\square$

Remark 7.20

The ambiguity in the definition of log presents no problem here. When we wrote $f(0) = 0$ we were taking the sensible view that $\log 1 = 0$. If, say, we insisted on taking $\log 1 = 2\pi i$ (which is certainly one of the values of $\mathrm{Log}\, 1$) then only the first term of the Taylor series would be altered.

Remark 7.21

The series for $1/(1+z)^2$ could be obtained from the series for $\log(1+z)$ by differentiating twice.

EXERCISES

7.9. Recall that f is an **even** function f if $f(-z) = f(z)$ for all z in $\mathbb{C}$, and that f is an **odd** function if $f(-z) = -f(z)$ for all z in $\mathbb{C}$. Let f be a holomorphic function with Taylor series

$$
f(z) = \sum_{n=0}^{\infty} a_n z^n \quad (|z| < R).
$$

Show that:

a) if f is even, then $a_n = 0$ for all odd n;

b) if f is odd, then $a_n = 0$ for all even n.

7.10. Let $c \in \mathbb{C}$. Determine Taylor series centred on c for e^z and $\cos z$.

7.11. Let f be an entire function, with Taylor series

$$f(z) = \sum_{n=0}^{\infty} a_n z^n .$$

For each $r > 0$, let $M(r) = \sup\{|f(z)| : z \in \kappa(0, r)\}$.

a) Deduce from (7.7) that

$$|a_n| \leq \frac{M(r)}{r^n} .$$

b) Suppose now that f is bounded. Give an alternative proof of Liouville's Theorem (Theorem 7.9) by deducing that $a_n = 0$ for all $n \geq 1$.

c) More generally, suppose that there exists $N \geq 1$ and $K > 0$ such that $|f(z)| \leq K|z^N|$ for all z. Show that f is a polynomial of degree at most N.

7.12. Let f, g be functions whose Taylor series

$$f(z) = \sum_{n=0}^{\infty} a_n z^n , \quad g(z) = \sum_{n=0}^{\infty} b_n z^n$$

have radii of convergence R_1, R_2, respectively. Let $h(z) = f(z)g(z)$, where $|z| < \min\{R_1, R_2\}$. By using Leibniz's formula for $h^{(n)}(z)$, show that h has Taylor series

$$h(z) = \sum_{n=0}^{\infty} c_n z^n ,$$

where $c_n = \sum_{r=0}^{n} a_{n-r} b_r$, the radius of convergence being at least $\min\{R_1, R_2\}$.

7.13. The odd function $\tan z$, being holomorphic in the open disc $N(0, \pi/2)$, has a Taylor series

$$\tan z = a_1 z + a_3 z^3 + a_5 z^5 + \cdots .$$

Use the identity $\sin z = \tan z \cos z$ and the result of the previous exercise to show that

$$\frac{(-1)^n}{(2n+1)!} = a_{2n+1} - \frac{a_{2n-1}}{2!} + \frac{a_{2n-3}}{4!} - \cdots + (-1)^n \frac{a_1}{(2n)!} \quad (n \geq 0).$$

Use this identity to calculate a_1, a_3 and a_5.

7.14. The odd function $\tanh z$ is also holomorphic in $N(0, \pi/2)$, and has a Taylor series

$$\tanh z = b_1 z + b_3 z^3 + b_5 z^5 + \cdots.$$

With the notation of the previous example, show that $b_{2n+1} = (-1)^n a_{2n+1}$.

8

Laurent Series and the Residue Theorem

8.1 Laurent Series

In Section 3.5 we looked briefly at functions with isolated singularities. It is clear that a function f with an isolated singularity at a point c cannot have a Taylor series centred on c. What it does have is a **Laurent**[1] **series**, a generalized version of a Taylor series in which there are negative as well as positive powers of $z - c$.

Theorem 8.1

Let f be holomorphic in the punctured disc $D'(c, R)$, where $R > 0$. Then there exist complex numbers a_n $(n \in \mathbb{Z})$ such that, for all z in $D'(a, R)$,

$$f(z) = \sum_{n=-\infty}^{\infty} a_n (z - c)^n .$$

If $0 < r < R$, then

$$a_n = \frac{1}{2\pi i} \int_{\kappa(c,r)} \frac{f(w)}{(w - c)^{n+1}} \, dw . \tag{8.1}$$

Proof

It will be sufficient to prove this for the case $c = 0$. Let $z \in D'(0, R)$, and let

---

[1] Pierre Alphonse Laurent, 1813–1854.

$0 < r_1 < |z| < r_2 < R.$

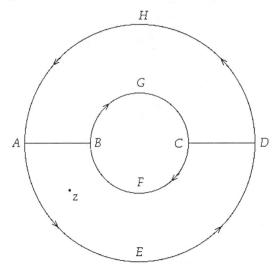

Let $z \in D(0, R)$. The function f is holomorphic inside and on both the contours

$$\gamma_1^* : A \to E \to D \to C \to F \to B \to A,$$
$$\gamma_2^* : A \to B \to G \to C \to D \to H \to A,$$

and we may suppose without loss of generality that z lies inside γ_1^*. Hence, since the function $f(w)/(w - z)$ is holomorphic on $\mathrm{I}(\gamma_2) \cup \gamma_2^*$,

$$\frac{1}{2\pi i} \int_{\gamma_2} \frac{f(w)}{w - z} \, dw = 0. \tag{8.2}$$

By contrast, it follows from Theorem 7.1 that

$$\frac{1}{2\pi i} \int_{\gamma_1} \frac{f(w)}{w - z} \, dw = f(z). \tag{8.3}$$

If we now add (8.2) and (8.3), the integrals along the straight line segments cancel each other. The outer and inner circles are traversed in the positive and negative directions respectively, and so

$$f(z) = \frac{1}{2\pi i} \int_{\kappa(0, r_2)} \frac{f(w)}{w - z} \, dw - \int_{\kappa(0, r_1)} \frac{f(w)}{w - z} \, dw. \tag{8.4}$$

For all w on the circle $\kappa(0, r_2)$ it is clear that $|z/w| < 1$. Thus

$$\frac{1}{w - z} = \frac{1}{w} \frac{1}{1 - (z/w)} = \frac{1}{w} \left(1 + (z/w) + (z/w)^2 + (z/w)^3 + \cdots \right)$$
$$= \sum_{n=0}^{\infty} \frac{z^n}{w^{n+1}}.$$

Similarly, $|w/z| < 1$ for all w on $\kappa(0, r_1)$, and so

$$\frac{1}{w-z} = -\sum_{n=0}^{\infty} \frac{w^n}{z^{n+1}} \, .$$

Hence, from (8.4),

$$f(z) = \frac{1}{2\pi i} \int_{\kappa(0,r_2)} f(w) \sum_{n=0}^{\infty} \frac{z^n}{w^{n+1}} \, dw + \frac{1}{2\pi i} \int_{\kappa(0,r_1)} f(w) \sum_{n=0}^{\infty} \frac{w^n}{z^{n+1}} \, dw \, . \quad (8.5)$$

Both the series are uniformly convergent by Theorem 5.33 and so, by Theorem 5.34,

$$f(z) = \sum_{n=0}^{\infty} a_n z^n + \sum_{n=0}^{\infty} b_n z^{-(n+1)} \, ,$$

where

$$a_n = \frac{1}{2\pi i} \int_{\kappa(0,r_2)} \frac{f(w)}{w^{n+1}} \, dw \, , \quad b_n = \frac{1}{2\pi i} \int_{\kappa(0,r_1)} f(w) w^n \, dw \, .$$

By the deformation theorem (Theorem 6.7) we can replace both $\kappa(0, r_1)$ and $\kappa(0, r_2)$ by $\kappa(0, r)$, where $0 < r < R$. Then, changing the notation by writing b_n as a_{-n-1}, we obtain the required result, that

$$f(z) = \sum_{n=-\infty}^{\infty} a_n z^n \, , \quad (8.6)$$

where

$$a_n = \frac{1}{2\pi i} \int_{\kappa(0,r)} \frac{f(w)}{w^{n+1}} \, dw \, . \quad (8.7)$$

$\square$

The series (8.6) is called the **Laurent expansion**, or **Laurent series** of f in the punctured disc $D'(0, R)$. The sum $g(z) = \sum_{n=-\infty}^{-1} a_n (z - c)^n$ is called the **principal part of f at c**.

There is a uniqueness theorem for Laurent series:

Theorem 8.2

Let f be holomorphic in the punctured disc $D'(c, R)$, and suppose that, for all z in $D'(c, R)$,

$$f(z) = \sum_{n=-\infty}^{\infty} b_n (z - c)^n \, . \quad (8.8)$$

Then, for all n in $\mathbb{Z}$,

$$b_n = \frac{1}{2\pi i} \int_{\kappa(c,r)} \frac{f(w)}{(w - c)^{n+1}} \, dw \, .$$

Proof

Again, it will be sufficient to consider the case where $c = 0$. Let a_n be as defined in (8.7), and let $r \in (0, R)$. Then

$$2\pi i a_n = \int_{\kappa(0,r)} \frac{f(w)}{w^{n+1}} \, dw = \int_{\kappa(0,r)} \left[\sum_{k=-\infty}^{\infty} b_k w^{k-n-1} \right] dw$$

$$= \int_{\kappa(0,r)} \left[\sum_{k=0}^{\infty} b_k w^{k-n-1} \right] dw + \int_{\kappa(0,r)} \left[\sum_{l=1}^{\infty} b_{-l} w^{-l-n-1} \right] dw \,.$$

Both these power series are convergent by assumption, and so, by Theorems 5.33 and 5.34, may be integrated term by term. Hence

$$2\pi i a_n = \sum_{k=-\infty}^{\infty} b_k \int_{\kappa(0,r)} w^{k-n-1} \, dw = 2\pi i b_n \,,$$

since, by Theorem 5.13, $\int_{\kappa(0,r)} w^{k-n-1} \, dw = 0$ unless $k - n - 1 = -1$. Thus $b_n = a_n$, and so the series (8.8) is indeed the Laurent expansion of f. $\qquad \square$

Corollary 8.3

If f has Laurent expansion

$$f(z) = \sum_{n=-\infty}^{\infty} a_n (z - c)^n$$

in the punctured disc $D'(c, R)$, then

$$\int_{\kappa(c,r)} f(z) \, dz = 2\pi i a_{-1} \,.$$

Proof

Simply put $n = -1$ in the formula (8.7). $\qquad \square$

This rather innocent result has far-reaching consequences, as we shall see shortly. The coefficient a_{-1} is called the **residue** of f at c, and we denote it by $\text{res}(f, c)$.

Example 8.4

Determine

$$\int_{\gamma} \frac{1}{z^2 + 1} \, dz \,,$$

where γ is the semicircle $[-R, R] \cup \{z : |z| = R, \text{Im } z \geq 0\}$, traversed in the positive direction, with $R > 1$.

Solution

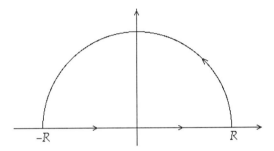

The integrand has a singularity at i. The Laurent expansion at i is given by

$$\frac{1}{z^2 + 1} = \frac{1}{(z-i)(z+i)} = \frac{1}{z-i} \cdot \frac{1}{2i + (z-i)} = \frac{1}{z-i} \frac{1}{2i} \left(1 + \frac{z-i}{2i}\right)^{-1}$$

$$= \frac{1}{2i(z-i)} \left(1 - \frac{z-i}{2i} + \frac{(z-i)^2}{(2i)^2} - \frac{(z-i)^3}{(2i)^3} + \cdots\right).$$

The function $z \mapsto 1/(z^2 + 1)$ has a simple pole at i, and the residue is $1/2i$. Hence

$$\int_\gamma \frac{1}{z^2 + 1} \, dz = 2\pi i(1/2i) = \pi.$$

□

Remark 8.5

You will notice that in this example nearly all of the Laurent expansion is irrelevant. We shall shortly consider techniques for obtaining the crucial coefficient a_{-1} without going to the trouble of finding the whole expansion.

The uniqueness theorem proves very useful in obtaining Laurent expansions, since the formula (8.7) for the coefficients often presents us with an integral that is far from easy to evaluate. For example, the function $\sin(1/z)$ is holomorphic in $\mathbb{C} \setminus \{0\}$, and the uniqueness theorem assures us that the obvious series

$$\sum_{n=0}^{\infty} (-1)^n \frac{z^{-2n-1}}{(2n+1)!}$$

is none other than the Laurent expansion. From (8.7) we note that

$$a_{-n} = \frac{1}{2\pi i} \int_{\kappa(0,r)} f(w)w^{n-1} \, dw,$$

and so we have the incidental conclusion that, for every non-negative integer n and every contour γ such that $0 \in I(\gamma)$,

$$\int_\gamma w^{2n} \sin(1/w)\, dw = \frac{(-1)^n 2\pi i}{(2n+1)!}\,,$$

while

$$\int_\gamma w^k \sin(1/w)\, dw = 0$$

unless k is even and non-negative. These conclusions would be hard to reach without the power of Laurent's Theorem.

Remark 8.6

For a function f that is holomorphic in an open domain containing the disc $D(c, R)$, we have a Taylor series

$$f(z) = \sum_{n=0}^{\infty} a_n (z - c)^n \,,$$

where

$$a_n = \frac{f^{(n)}(c)}{n!} = \frac{1}{2\pi i} \int_{\kappa(c,R)} \frac{f(w)}{(w - c)^{n+1}}\, dw\,.$$

This is also the Laurent expansion. If f is holomorphic throughout the disc then the negative coefficients in the Laurent expansion are all 0.

It is often sufficient to know the first few terms of a Laurent expansion, and here the O and o notations can save a lot of unnecessary detail.

Example 8.7

Calculate the first few terms of the Laurent series for $1/\sin z$ at 0.

Solution

The function $1/\sin z$ has a singularity at 0 but is otherwise holomorphic in the neighbourhood $N(0, \pi)$. We know that, as $z \to 0$,

$$\sin z = z - \frac{z^3}{6} + O(z^5)\,.$$

Hence, for z near 0,

$$\frac{1}{\sin z} = \frac{1}{z}\left[1 - \left(\frac{z^2}{6} + O(z^4)\right)\right]^{-1} = \frac{1}{z}\left[1 + \frac{z^2}{6} + O(z^4)\right]$$

$$= \frac{1}{z} + \frac{z}{6} + O(z^3)\,.$$

□

If we require more terms it is in principle easy to compute them. See Exercise 8.1.

Example 8.8

Show that

$$\cot z = \frac{1}{z} - \frac{z}{3} + O(z^3).$$

Solution

From what we know already,

$$\cot z = \frac{\cos z}{\sin z} = \left(1 - \frac{z^2}{2} + O(z^4)\right)\left(\frac{1}{z} + \frac{z}{6} + O(z^3)\right)$$

$$= \frac{1}{z} - z\left(\frac{1}{2} - \frac{1}{6}\right) + O(z^3)$$

$$= \frac{1}{z} - \frac{z}{3} + O(z^3).$$

□

EXERCISES

8.1. Show that

$$\frac{1}{\sin z} = \frac{1}{z} + \frac{z}{6} + \frac{7z^3}{360} + o(z^4).$$

8.2. Show that

$$\frac{1}{1 - \cos z} = \frac{2}{z^2} + \frac{1}{6} + \frac{z^2}{120} + o(z^3).$$

8.3. Show that the coefficient of z^{-1} in the Laurent series of $e^{1/z}e^{2z}$ is

$$\sum_{n=0}^{\infty} \frac{2^n}{n!(n+1)!}.$$

8.4. Determine $\operatorname{res}(f, 0)$, where:

a) $f(z) = 1/(z^4 \sin z)$;

b) $f(z) = 1/[z^3(1 - \cos z)]$.

8.2 Classification of Singularities

We encountered singularities in Section 3.5, but the Laurent series helps us to understand them better. Let $f(z)$ have a Laurent series $\sum_{n=-\infty}^{\infty} a_n(z-c)^n$. If $a_n \neq 0$ for infinitely many negative values of n, then f has an **essential singularity** at c. Otherwise, if n is the least integer (positive or negative) such that $a_n \neq 0$, we say that n is the **order** of f at c, and write $n = \mathrm{ord}(f, c)$. It is clear that $\mathrm{ord}(f, c)$ is the unique integer n such that $f(z) = (z-c)^n g(z)$, where g is differentiable and non-zero at c.

Example 8.9

Show that $\cos(1/z)$ has an essential singularity at 0.

Solution

Since $\cos(1/z)$ has the Laurent series

$$1 - \frac{1}{2!}z^{-2} + \frac{1}{4!}z^{-4} - \cdots,$$

it is clear that it has an essential singularity at 0.   □

The coefficients of the Laurent series for f do not depend on the value of $f(c)$, which we may take as undefined. If $\mathrm{ord}(f, c) = n \geq 0$ then f becomes differentiable at c if we define $f(c)$ as a_0. Thus, whether $f(c)$ was undefined, or had a value other than a_0, we can remove the singularity at c by defining (or redefining) $f(c)$ to be a_0. This is what is called a **removable singularity**. Note that f has (at worst) a removable singularity at c if and only if $\lim_{z \to c} f(z)$ is finite.

If $\mathrm{ord}(f, c) = -m$, where $m > 0$, then f has a **pole of order** m at c. A pole of order 1 is usually called a **simple pole**. From Examples 8.7 and 8.8 we see that sin and cot both have simple poles at 0, and from Exercise 7.2 we see that $1/(1 - \cos z)$ has a pole of order 2 (a **double pole**) at 0.

It is clear that if c is a pole of f then $f(z) \to \infty$ as $z \to c$. If f has an essential singularity at c, then $\lim_{z \to c} f(z)$ does not exist. Indeed we have the following remarkable theorem, due to Casorati[2] and Weierstrass, which says that within an arbitrarily small punctured disc $D'(c, \delta)$ the value of $f(z)$ can be made arbitrarily close to any complex number whatever:

---

[2] Felice Casorati, 1835–1890.

Theorem 8.10 (The Casorati–Weierstrass Theorem)

Let f have an essential singularity at c, and let d be an arbitrary complex number. Then, for all $\delta > 0$ and for all $\epsilon > 0$ there exists z in $D'(c, \delta)$ such that $|f(z) - d| < \epsilon$.

Proof

Suppose, for a contradiction, that for some d in $\mathbb{C}$ there exists $\epsilon > 0$ and $\delta > 0$ such that $|f(z) - d| \geq \epsilon$ for all z in $D'(c, \delta)$. Let $g(z) = 1/\big(f(z) - d\big)$. Then, for all z in $D'(c, \delta)$,

$$|g(z)| \leq \frac{1}{\epsilon}$$

and so g has (at worst) a removable singularity at c. Since g is not identically zero, $\mathrm{ord}(g, c) = k \geq 0$, and so

$$\mathrm{ord}(f, c) = \mathrm{ord}(f - d, c) = -k \,.$$

(See Exercise 8.5 below.) This contradicts the assumption that f has an essential singularity at c.   □

As a consequence we have the following result, which says that a non-polynomial entire function comes arbitrarily close to every complex number in any region $\{z : |z| > R\}$:

Theorem 8.11

Let f be an entire function, not a polynomial. Let $R > 0$, $\epsilon > 0$ and $c \in \mathbb{C}$. Then there exists z such that $|z| > R$ and $|f(z) - c| < \epsilon$.

Proof

The function f has a non-terminating Taylor series $\sum_{n=0}^{\infty} a_n z^n$, converging for all z. It follows that the function g, defined by $g(z) = f(1/z)$, has an essential singularity at 0. So, by Theorem 8.10, for all $\epsilon, R > 0$ and all c in $\mathbb{C}$, there exists z such that $|z| < 1/R$ and $|g(z) - c| < \epsilon$. That is, $|1/z| > R$ and $|f(1/z) - c| < \epsilon$.   □

EXERCISES

8.5. Let f and g have finite order at c. Show that:

a) $\operatorname{ord}(f \cdot g, c) = \operatorname{ord}(f, c) + \operatorname{ord}(g, c)$;

b) $\operatorname{ord}(1/f,\, c) = -\operatorname{ord}(f, c)$;

c) if $\operatorname{ord}(f, c) < \operatorname{ord}(g, c)$, then $\operatorname{ord}(f + g, c) = \operatorname{ord}(f, c)$.

8.6 Use the previous exercise to deduce that

a) $1/z \sin^2 z$ has a triple pole at 0;

b) $(\cot z + \cos z)/\sin 2z$ has a double pole at 0.

c) $z^2(z - 1)/[(1 - \cos z)\log(1 + z)]$ has a simple pole at zero.

8.3 The Residue Theorem

Let γ be a contour and let f be holomorphic in a domain containing $\mathrm{I}(\gamma) \cup \gamma^*$, except for a single point c in $\mathrm{I}(\gamma)$. Then f has a Laurent expansion

$$f(z) = \sum_{n=-\infty}^{\infty} a_n (z - c)^n \,,$$

valid for all $z \neq c$ in $\mathrm{I}(\gamma) \cup \gamma^*$. From Corollary 8.3 and the Deformation Theorem (Theorem 6.7), we deduce that

$$\int_\gamma f(z)\, dz = 2\pi i a_{-1} \,.$$

We refer to a_{-1} as the **residue** of f at the singularity c, and write

$$a_{-1} = \operatorname{res}(f, c) \,. \tag{8.9}$$

The next result extends this conclusion to a function having finitely many singularities within the contour:

Theorem 8.12 (The Residue Theorem)

Let γ be a contour, and let f be a function holomorphic in an open domain U containing $\mathrm{I}(\gamma) \cup \gamma^*$, except for finitely many poles at $c_1, c_2, \ldots, c_m$ in $\mathrm{I}(\gamma)$. Then

$$\int_\gamma f(z)\, dz = 2\pi i \sum_{k=1}^{m} \operatorname{res}(f, c_k) \,.$$

Proof

For $k = 1, 2, \ldots, m$, let f_k be the principal part of f at c_k. Suppose in fact that c_k is a pole of order N_k, so that the Laurent series of f at c_k is

$$\sum_{n=-N_k}^{\infty} a_n^{(k)}(z - c_k)^n \,.$$

Then $f_k(z) = \sum_{n=-N_k}^{-1} a_n^{(k)}(z - c_k)^n$, a rational function with precisely one singularity, a pole of order N_k at c_k. Notice also that

$$f(z) - f_k(z) = \sum_{n=0}^{\infty} a_n^{(k)}(z - c_k)^n$$

and so $f - f_k$ is holomorphic in some neighbourhood of c_k.

Let $g = f - (f_1 + f_2 + \cdots + f_m)$. We write $g = (f - f_k) - \sum_{j \neq k} f_j$ and observe that $f - f_k$ and each f_j $(j \neq k)$ are holomorphic at c_k. This happens for each value of k and, since there are no other potential singularities for g, we conclude that g is holomorphic in U. Hence $\int_\gamma g(z)\,dz = 0$, and so

$$\int_\gamma f(z)\,dz = \sum_{k=1}^{m} \int_\gamma f_k(z)\,dz\,. \tag{8.10}$$

By Theorem 5.13,

$$\int_\gamma f_k(z)\,dz = 2\pi i a_{-1}^{(k)} = 2\pi i \operatorname{res}(f, c_k)\,,$$

and the result now follows immediately from (8.10). $\qquad\square$

Accordingly, the key to integration round a contour is the calculation of residues, and it is important to be able to calculate those without computing the entire Laurent series. Simple poles are the easiest:

Theorem 8.13

Let f have a simple pole at c. Then

$$\operatorname{res}(f, c) = \lim_{z \to c}(z - c)f(z)\,.$$

Thus, if

$$f(z) = \frac{a}{z - c} + O(1)\,,$$

then $\operatorname{res}(f, c) = a$.

Proof

Suppose that f has a simple pole at c, so that the Laurent series is

$$f(z) = a_{-1}(z - c)^{-1} + \sum_{n=0}^{\infty} a_n(z - c)^n.$$

Then, as $z \to c$,

$$(z - c)f(z) = a_{-1} + \sum_{n=0}^{\infty} a_n(z - c)^{n+1} \to a_{-1} = \mathrm{res}(f, c).$$

$\square$

Example 8.14

Evaluate

$$\int_\gamma \frac{\sin(\pi z)}{z^2 + 1}\, dz$$

where γ is any contour such that $i, -i \in \mathrm{I}(\gamma)$.

Solution

The integrand f has simple poles at i and $-i$. Recalling that $\sin(iz) = i \sinh z$, we obtain from Theorem 8.13 that

$$\mathrm{res}(f, i) = \lim_{z \to i} \frac{\sin(\pi z)}{z + i} = \frac{\sinh \pi}{2},$$

$$\mathrm{res}(f, -i) = \lim_{z \to -i} \frac{\sin(\pi z)}{z - i} = \frac{\sinh \pi}{2}.$$

Hence

$$\int_\gamma \frac{\sin(\pi z)}{z^2 + 1}\, dz = i\pi \sinh \pi.$$

$\square$

In that example, and in many others, the integrand $f(z)$ is of the form $g(z)/h(z)$, where both g and h are holomorphic, and where $h(c) = 0$, $h'(c) \neq 0$. A technique applying to this situation is worth recording as a theorem:

Theorem 8.15

Let $f(z) = g(z)/h(z)$, where g and h are both holomorphic in a neighbourhood of c, and where $h(c) = 0$, $h'(c) \neq 0$. Then

$$\mathrm{res}(f, c) = \frac{g(c)}{h'(c)}.$$

Proof

$$\operatorname{res}(f,c) = \lim_{z \to c} g(z) \frac{z-c}{h(z)} = g(c) \lim_{z \to c} \frac{z-c}{h(z) - h(c)} = \frac{g(c)}{h'(c)}.$$

$\square$

In the last example this observation makes little or no difference, but it can help in other cases.

Example 8.16

Evaluate

$$\int_\gamma \frac{dz}{z^4 + 1},$$

where γ is the semicircle $[-R, R] \cup \{z : |z| = R \text{ and } \operatorname{Im} z > 0\}$, traced in the positive direction, and $R > 1$.

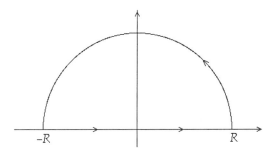

Solution

Within the semicircle, the integrand f has two simple poles, at $e^{i\pi/4}$ and $e^{3i\pi/4}$. From (8.15) we see that

$$\operatorname{res}(f, e^{i\pi/4}) = \frac{1}{4e^{3i\pi/4}} = \frac{e^{-3i\pi/4}}{4} = \frac{-1-i}{4\sqrt{2}},$$

$$\operatorname{res}(f, e^{3i\pi/4}) = \frac{1}{4e^{9i\pi/4}} = \frac{e^{-i\pi/4}}{4} = \frac{1-i}{4\sqrt{2}}.$$

Hence

$$\int_\gamma \frac{dz}{z^4 + 1} = 2\pi i \frac{-2i}{4\sqrt{2}} = \frac{\pi}{\sqrt{2}}.$$

$\square$

Multiple poles are a little more troublesome. A more general version of Theorem 8.13 is available, but it does not always give the most effective way of computing the residue:

Theorem 8.17

Suppose that f has a pole of order m at c, so that

$$f(z) = \sum_{n=-m}^{\infty} a_n (z - c)^n,$$

and $a_{-m} \neq 0$. Then

$$\operatorname{res}(f, c) = \frac{1}{(m-1)!}\, g^{(m-1)}(c),$$

where $g(z) = (z - c)^m f(z)$.

Proof

From

$$g(z) = a_{-m} + a_{m+1}(z - c) + \cdots + a_{-1}(z - c)^{m-1} + \cdots$$

we deduce, differentiating $m - 1$ times, that

$$g^{(m-1)}(z) = (m-1)!a_{-1} + \text{ positive powers of } (z - c);$$

hence

$$g^{(m-1)}(c) = (m-1)!a_{-1} = (m-1)!\operatorname{res}(f, c).$$

$\square$

Example 8.18

Evaluate

$$\int_\gamma \frac{dz}{(z^2 + 1)^2},$$

where γ is the semicircle $[-R, R] \cup \{z : |z| = R \text{ and } \operatorname{Im} z > 0\}$, traced in the positive direction, and $R > 1$.

Solution

The integrand f has a double pole at i, and $f(z) = (z - i)^{-2}g(z)$, where $g(z) = 1/(z + i)^2$. Hence $g'(z) = -2/(z + i)^3$, and so

$$\operatorname{res}(f, i) = g'(i) = \frac{-2}{(2i)^3} = \frac{-2}{-8i} = \frac{1}{4i}.$$

Hence

$$\int_\gamma \frac{dz}{(z^2 + 1)^2} = 2\pi i \,\frac{1}{4i} = \frac{\pi}{2}.$$

$\square$

For multiple poles it is frequently less troublesome to calculate the relevant terms of the Laurent series.

Example 8.19

Find the residue of $1/(z^2 \sin z)$ at the triple pole 0.

Solution

$$\frac{1}{z^2 \sin z} = \frac{1}{z^2}\left(z - \frac{z^3}{6} + O(z^5)\right)^{-1} = \frac{1}{z^3}\left(1 - \frac{z^2}{6} + O(z^4)\right)^{-1}$$
$$= \frac{1}{z^3}\left(1 + \frac{z^2}{6} + O(z^4)\right).$$

The residue is the coefficient of z^{-1}, namely $1/6$.

The alternative method, which involves calculating $\lim_{z\to 0} g''(z)$, where $g(z) = z^3 f(z) = z/\sin z$, is much harder.   □

EXERCISES

8.7. Let f be an **even** meromorphic function, that is to say, let f be such that $f(-z) = f(z)$ for all z, and suppose that f has a pole at 0. Show that $\operatorname{res}(f, 0) = 0$.

8.8. Calculate the residue of $1/(z - \sin z)$ at 0.

8.9. Calculate the residue of $1/(z^3 + 1)^2$ at -1.

8.10. Calculate the residue at 0 of
$$\frac{1 + z^6}{z^3(1 - 2z)(2 - z)}.$$

8.11. Using the method of Example 8.19, show that the residue of $\cot \pi z/z^2$ at the triple pole 0 is $-\pi/3$.

8.12. Show that $\cot \pi z$ and $\operatorname{cosec} \pi z$ have simple poles at every integer n, and that
$$\operatorname{res}(\cot \pi z, n) = \frac{1}{\pi}, \quad \operatorname{res}(\operatorname{cosec} \pi z, n) = \frac{(-1)^n}{\pi}.$$

Show, more generally, that, if f has no zeros on the x-axis,
$$\operatorname{res}(\pi f(z) \cot \pi z, n) = f(n), \quad \operatorname{res}(\pi f(z) \operatorname{cosec} \pi z, n) = (-1)^n f(n).$$

8.13. Evaluate the integrals:

a)

$$\int_{\kappa(0,2)} \frac{\sin \pi z \, dz}{(2z + 1)^3} \, .$$

b)

$$\int_{\kappa(0,\pi/4)} \frac{dz}{z^2 \tan z} \, .$$

9

Applications of Contour Integration

9.1 Real Integrals: Semicircular Contours

One of the very attractive features of complex analysis is that it can provide elegant and easy proofs of results in real analysis. Let us look again at Example 8.16. The contour γ is parametrised by

$$\left.\begin{array}{ll} \gamma_1(t) = t + i0 & (-R \le t \le R) \\ \gamma_2(t) = Re^{it} & (0 < t < \pi), \end{array}\right\} \tag{9.1}$$

and so

$$\frac{\pi}{\sqrt{2}} = \int_\gamma \frac{dz}{z^4+1} = \int_{-R}^{R} \frac{dt}{t^4+1} + \int_0^\pi \frac{iRe^{it}}{R^4 e^{4it}+1}\, dt. \tag{9.2}$$

By Theorem 5.24,

$$\left| \int_0^\pi \frac{iRe^{it}}{R^4 e^{4it}+1}\, dt \right| \le \frac{\pi R}{R^4-1},$$

which tends to 0 as $R \to \infty$. Hence, letting $R \to \infty$ in (9.2), we obtain the Cauchy principal value:

$$(PV) \int_{-\infty}^{\infty} \frac{dt}{t^4+1} = \frac{\pi}{\sqrt{2}}.$$

Since

$$\frac{1}{t^4+1} \sim \frac{1}{t^4} \text{ as } t \to \pm\infty,$$

both

$$\int_0^\infty \frac{dt}{t^4+1} \quad \text{and} \quad \int_{-\infty}^0 \frac{dt}{t^4+1}$$

converge. So we can leave out the (PV) and conclude that

$$\int_{-\infty}^{\infty} \frac{dt}{t^4 + 1} = \frac{\pi}{\sqrt{2}}.$$

If necessary, we may then deduce by symmetry that

$$\int_{0}^{\infty} \frac{dt}{t^4 + 1} = \frac{\pi}{2\sqrt{2}}.$$

This result can of course be obtained by elementary methods, but the process is lengthy and tedious.

Similarly, if we examine Example 8.18, in which the contour has the same parametrisation (9.1), we find that

$$\frac{\pi}{2} = \int_{\gamma} \frac{dz}{(z^2 + 1)^2} = \int_{-R}^{R} \frac{dt}{(t^2 + 1)^2} + \int_{0}^{\pi} \frac{iRe^{it}\, dt}{(R^2 e^{2it} + 1)^2},$$

and

$$\left| \int_{0}^{\pi} \frac{iRe^{it}\, dt}{(R^2 e^{2it} + 1)^2} \right| \leq \frac{\pi R}{(R^2 - 1)^2},$$

which again tends to 0 as $R \to \infty$. Hence

$$\int_{-\infty}^{\infty} \frac{dt}{(t^2 + 1)^2} = \frac{\pi}{2},$$

since, as in the last example, we can dispense with the (PV) in front of the integral. Here the elementary method is not quite so tedious as in the previous case: substituting $t = \tan\theta$ gives

$$\int_{-\infty}^{\infty} \frac{dt}{(t^2 + 1)^2} = \int_{-\pi/2}^{\pi/2} \frac{\sec^2\theta}{\sec^4\theta}\, d\theta = \int_{-\pi/2}^{\pi/2} \cos^2\theta\, d\theta$$

$$= \frac{1}{2} \int_{-\pi/2}^{\pi/2} (1 + \cos 2\theta)\, d\theta = \frac{1}{2} \left[\theta + \tfrac{1}{2} \sin 2\theta \right]_{-\pi/2}^{\pi/2} = \frac{\pi}{2}.$$

The two integrals we have examined are both special cases of the following general theorem.

Theorem 9.1

Let f be a complex function with the properties:

(i) f is meromorphic in the upper half-plane, having poles at finitely many points $p_1, p_2, \ldots, p_n$;

(ii) f has no poles on the real axis;

(iii) $zf(z) \to 0$ uniformly in the upper half-plane, as $|z| \to \infty$;

(iv) $\int_0^\infty f(x)\,dx$ and $\int_{-\infty}^0 f(x)\,dx$ both converge.

Then

$$\int_{-\infty}^\infty f(x)\,dx = 2\pi i \sum_{k=1}^n \operatorname{res}(f, p_k)\,.$$

Proof

Consider $\int_\gamma f(z)\,dz$, where the contour γ is the semicircle of radius R in the upper half-plane. Thus

$$\int_\gamma f(z)\,dz = \int_{-R}^R f(x)\,dx + \int_0^\pi f(Re^{i\theta})iRe^{i\theta}\,d\theta\,. \tag{9.3}$$

From Condition (iii) we know that, for all $\epsilon > 0$, there exists $K > 0$ such that $|zf(z)| < \epsilon$ for all z in the upper half-plane such that $|z| > K$. Hence, for all $R > K$,

$$\left| \int_0^\pi f(Re^{i\theta})iRe^{i\theta}\,d\theta \right| < \pi\epsilon\,.$$

Thus $\int_0^\pi f(Re^{i\theta})iRe^{i\theta}\,d\theta \to 0$ as $R \to \infty$ and so, letting R tend to ∞ in (9.3) and applying the residue theorem, we see that

$$(PV)\int_{-\infty}^\infty f(x)\,dx = 2\pi i \sum_{k=1}^n \operatorname{res}(f, p_k)\,.$$

Condition (iv) means that we can dispense with the principal value prefix (PV). $\qquad\square$

Example 9.2

Evaluate

$$\int_{-\infty}^\infty \frac{\cos x\,dx}{x^2 + 2x + 4}\,.$$

Solution

Let

$$f(z) = \frac{e^{iz}}{z^2 + 2z + 4}\,.$$

Then $f(z)$ has two poles, at $-1 + i\sqrt{3}$ and $-1 - i\sqrt{3}$, and only the first of these is in the upper half-plane. From (8.15) we calculate that

$$\operatorname{res}(f, -1 + i\sqrt{3}) = \frac{e^{i(-1+i\sqrt{3})}}{2(-1 + i\sqrt{3}) + 2} = \frac{e^{-i}e^{-\sqrt{3}}}{2i\sqrt{3}}\,.$$

In the upper half-plane, for all sufficiently large $|z|$,

$$|zf(z)| = \left| \frac{ze^{iz}}{z^2 + 2z + 4} \right|$$

$$= \left| \frac{ze^{ix}e^{-y}}{z^2 + 2z + 4} \right| \quad \text{(where } x = \operatorname{Re} z \text{ and } y = \operatorname{Im} z\text{)}$$

$$\leq \frac{|z|}{|z|^2 - 2|z| - 4} \quad \text{(since } |e^{ix}| = 1 \text{ and } |e^{-y}| < 1\text{)}.$$

It follows that $|zf(z)|$ tends uniformly to 0 in the upper half plane.

All the conditions of Theorem 9.1 are satisfied, and so

$$\int_{-\infty}^{\infty} \frac{e^{ix}}{x^2 + 2x + 4} = 2\pi i \operatorname{res}(f, -1 + i\sqrt{3}) = \frac{\pi}{\sqrt{3}} e^{-i} e^{-\sqrt{3}}.$$

Equating real parts gives

$$\int_{-\infty}^{\infty} \frac{\cos x \, dx}{x^2 + 2x + 4} = \frac{\pi}{\sqrt{3}} e^{-\sqrt{3}} \cos 1.$$

Our method gives us a bonus, since equating imaginary parts gives

$$\int_{-\infty}^{\infty} \frac{\sin x \, dx}{x^2 + 2x + 4} = -\frac{\pi}{\sqrt{3}} e^{-\sqrt{3}} \sin 1.$$

$\square$

Remark 9.3

An approach using

$$f(z) = \frac{\cos z}{z^2 + 2z + 4}$$

would not work, since $\cos iy = \cosh y \sim \frac{1}{2}e^y$ as $y \to \infty$, and so Condition (iii) certainly fails.

EXERCISES

9.1 Show that

$$\int_0^{\infty} \frac{x^2}{1 + x^4} \, dx = \frac{\pi}{2\sqrt{2}}.$$

9.2 Show that

$$\int_{-\infty}^{\infty} \frac{x^2}{(x^2 + 1)(x^2 + 4)} \, dx = \frac{\pi}{3}.$$

9.3. Show that
$$\int_0^\infty \frac{1}{x^6 + 1} \, dx = \frac{\pi}{3} \, .$$

9.4. Show that
$$\int_{-\infty}^\infty \frac{1}{(x^2 + 1)^3} \, dx = \frac{3\pi}{8} \, .$$

9.5. More generally, show that
$$\int_{-\infty}^\infty \frac{1}{(x^2 + 1)^n} \, dx = \frac{\pi}{2^{2n-2}} \binom{2n - 2}{n - 1} \, .$$

9.6. Show that
$$\int_{-\infty}^\infty \frac{1}{(x^2 + x + 1)^2} \, dx = \frac{4\pi}{3\sqrt{3}} \, .$$

9.7. Show that
$$\int_0^\infty \frac{\cos x}{x^2 + 1} \, dx = \frac{\pi}{2e} \, .$$

9.8. By expressing $\sin^2 x$ as $\frac{1}{2}(1 - \cos 2x)$, determine the value of
$$\int_0^\infty \frac{\sin^2 x \, dx}{1 + x^2} \, .$$

9.9. Suppose that $c > 0$, $d > 0$ and $c \neq d$. Show that
$$\int_{-\infty}^\infty \frac{\cos x \, dx}{(x^2 + c^2)(x^2 + d^2)} = \frac{\pi}{c^2 - d^2} \left(\frac{e^{-d}}{d} - \frac{e^{-c}}{c} \right) \, .$$

9.10. Show that, if $c > 0$,
$$\int_{-\infty}^\infty \frac{\cos x \, dx}{(x^2 + c^2)^2} = \frac{\pi(c + 1)e^{-c}}{2c^3} \, .$$

9.11. Show that, if $k > 0$ and $s > 0$,
$$\int_{-\infty}^\infty \frac{\cos sx}{k^2 + z^2} \, dz = \frac{\pi}{k} e^{-ks} \, .$$

9.2 Integrals Involving Circular Functions

We are familiar with the idea of rewriting an integral

$$\int_{\kappa(0,1)} f(z)\, dz$$

round the unit circle $\kappa(0,1)$ as an integral

$$\int_0^{2\pi} f(e^{i\theta}) i e^{i\theta}\, d\theta \,.$$

Sometimes it is useful to reverse the process. Consider an integral of the form

$$\int_0^{2\pi} f(\sin\theta, \cos\theta)\, d\theta \,.$$

If $z = e^{i\theta}$ then

$$\cos\theta = \frac{1}{2}\left(z + \frac{1}{z}\right)\,, \quad \sin\theta = \frac{1}{2i}\left(z - \frac{1}{z}\right)\,, d\theta = \frac{dz}{iz}\,,$$

and so the integral can be expressed as $\int_{\kappa(0,1)} g(z)\, dz$.

Some examples will demonstrate the technique.

Example 9.4

Evaluate

$$I = \int_0^{2\pi} \frac{d\theta}{a + b\cos\theta} \quad (a > b > 0)\,.$$

Solution

The substitution gives

$$I = \int_{\kappa(0,1)} \frac{1}{a + \frac{1}{2}b(z + z^{-1})}\, \frac{dz}{iz} = -2i \int_{\kappa(0,1)} \frac{dz}{bz^2 + 2az + b}$$

$$= -2i \int_{\kappa(0,1)} \frac{dz}{b(z - \alpha)(z - \beta)}\,,$$

where

$$\alpha = \frac{-a + \sqrt{a^2 - b^2}}{b} \quad \text{and} \quad \beta = \frac{-a - \sqrt{a^2 - b^2}}{b}$$

are the roots of the equation $bz^2 + 2az + b = 0$. The integrand has simple poles at α and β. Now,

$$|\beta| = \frac{a + \sqrt{a^2 - b^2}}{b} > \frac{a}{b} > 1\,,$$

and so β does not lie within the contour $\kappa(0,1)$. Since $\alpha\beta = 1$ we deduce that $|\alpha| < 1$. The residue at α is

$$\frac{1}{b(\alpha - \beta)} = \frac{1}{2\sqrt{a^2 - b^2}},$$

and so

$$I = 2\pi i(-2i)\frac{1}{2\sqrt{a^2 - b^2}} = \frac{2\pi}{\sqrt{a^2 - b^2}}.$$

□

The technique requires that the limits of integration be 0 and 2π. It is, however, sometimes possible to transform an integral into the required form.

Example 9.5

Evaluate

$$I = \int_0^\pi \frac{d\theta}{a^2 + \cos^2\theta} \quad (a > 0).$$

Solution

We use the identity $2\cos^2\theta = 1 + \cos 2\theta$ and then substitute $\phi = 2\theta$ to obtain

$$I = \int_0^{2\pi} \frac{d\phi}{(2a^2 + 1) + \cos\phi}.$$

Then from the previous example we deduce that

$$I = \frac{\pi}{a\sqrt{a^2 + 1}}.$$

□

Example 9.6

Evaluate

$$I = \int_0^{2\pi} \frac{\cos\theta\, d\theta}{1 - 2a\cos\theta + a^2} \quad (|a| < 1).$$

Solution

With $z = e^{i\theta}$ as usual, we find that

$$I = \frac{1}{2i}\int_{\kappa(0,1)} \frac{1 + z^2}{z(1 - az)(z - a)}\, dz.$$

The integrand has simple poles at 0, a and $1/a$, the first two of which lie inside the contour. The residue at 0 is $-1/a$, and at a is $(1+a^2)/[a(1-a^2)]$. Hence

$$I = \pi \left(\frac{1+a^2}{a(1-a^2)} - \frac{1}{a} \right) = \frac{\pi(1+a^2-1+a^2)}{a(1-a^2)} = \frac{2\pi a}{1-a^2}.$$

□

Sometimes it is easier to solve the problem by considering the integral from 0 to 2π of a complex-valued function. The next example could be done by substituting $(z^3 + z^{-3})/2$ for $\cos 3\theta$, but the resulting integral is much more difficult to evaluate.

Example 9.7

Show that

$$\int_0^{2\pi} \frac{\cos 3\theta \, d\theta}{5 - 4\cos\theta} = \frac{\pi}{12}.$$

Solution

$$\int_0^{2\pi} \frac{e^{3i\theta} \, d\theta}{5 - 4\cos\theta} = \int_{\kappa(0,1)} \frac{z^3 \, dz}{iz[5 - 2(z + z^{-1})]} = i \int_{\kappa(0,1)} \frac{z^3 \, dz}{2z^2 - 5z + 2}$$

$$= i \int_{\kappa(0,1)} \frac{z^3}{(2z-1)(z-2)} \, dz.$$

The integrand has a simple pole within $\kappa(0,1)$ at $1/2$, and the residue is $-1/24$. Hence, equating real parts, we deduce that

$$\int_0^{2\pi} \frac{\cos 3\theta \, d\theta}{5 - 4\cos\theta} = \text{Re}[2\pi i \,.\, i \,.\, (-1/24)] = \frac{\pi}{12}.$$

□

EXERCISES

9.12. Evaluate

$$\int_0^{2\pi} \frac{d\theta}{1 - 2a\cos\theta + a^2}$$

(i) if $|a| < 1$; (ii) if $|a| > 1$.

9.13. Show that

$$\int_0^{2\pi} \frac{\cos^2 2\theta \, d\theta}{1 - 2p\cos\theta + p^2} = \frac{\pi(1+p^4)}{1-p^2} \quad (0 < p < 1).$$

9.14. Show that, if $a > b > 0$,

$$\int_0^{2\pi} \frac{\sin^2\theta\, d\theta}{a + b\cos\theta} = \frac{2\pi}{b^2}\left(a - \sqrt{a^2 - b^2}\,\right).$$

9.15. Let $\alpha > 0$. Show that

$$\int_0^{2\pi} \frac{\cos 3\theta\, d\theta}{\cosh\alpha - \cos\theta} = \frac{2\pi e^{-3\alpha}}{\sinh\alpha}\,.$$

By putting $\alpha = \cosh^{-1}(5/4)$, recover the result of Example 9.7.

9.16. Evaluate

$$\int_{\kappa(0,1)} \frac{e^z}{z^{n+1}}\, dz\,.$$

Deduce that

$$\int_0^{2\pi} e^{\cos\theta}\cos(n\theta - \sin\theta)\, d\theta = \frac{2\pi}{n!}\,.$$

9.17. Let $m \in \mathbb{R}$ and $a \in (-1,1)$. By integrating $e^{mz}/(z - ia)$ round the circle $\kappa(0,1)$, show that

$$\int_0^{2\pi} \frac{e^{m\cos\theta}[\cos(m\sin\theta) - a\sin(m\sin\theta + \theta)]}{1 - 2\sin\theta + a^2}\, d\theta = 2\pi\cos ma\,,$$

$$\int_0^{2\pi} \frac{e^{m\cos\theta}[\sin(m\sin\theta) + a\cos(m\sin\theta + \theta)]}{1 - 2\sin\theta + a^2}\, d\theta = 2\pi\sin ma\,.$$

9.3 Real Integrals: Jordan's Lemma

The range of applications to real integrals is extended by the use of the following result, usually known as Jordan's Lemma:

Theorem 9.8 (Jordan's Lemma)

Let f be differentiable in $\mathbb{C}$, except at finitely many poles, none of which lies on the real line, and let $c_1, c_2, \ldots, c_n$ be the poles in the upper half-plane. Suppose also that $f(z) \to 0$ as $z \to \infty$ in the upper half-plane. Then, for every positive real number a,

$$\int_{-\infty}^{\infty} f(x)\cos ax\, dx + i\int_{-\infty}^{\infty} f(x)\sin ax\, dx = 2\pi i\sum_{k=1}^{n} \mathrm{res}(g, c_i)\,,$$

where $g(z) = f(z)e^{iaz}$.

Proof

Let $\epsilon > 0$. Let R be such that

(i) $|c_k| < R$ for $k = 1, 2, \ldots, n$;

(ii) $|f(z)| \le \epsilon$ for all z such that $|z| \ge R$ and $\operatorname{Im} z > 0$;

(iii) $xe^{-ax} \le 1$ for all $x \ge R$.

Let $u, v > R$. Let σ be the square with vertices $-u$, v, $v + iw$, $-u + iw$, where $w = u + v$.

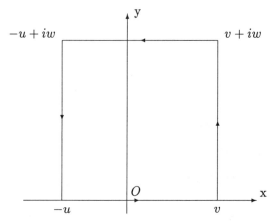

From the Residue Theorem (Theorem 8.12) we know that

$$\int_\sigma g(z)\,dz = 2\pi i \sum_{k=1}^n \operatorname{res}(g, c_k) \,,$$

for the poles of g are in the same places as those of f. Now,

$$\int_\sigma g(z)\,dz = \int_{-u}^v f(x)e^{iax}\,dx + \int_0^w f(v + iy)e^{iav - ay}\,dy$$
$$- \int_{-u}^v f(x + iw)e^{iax - aw}\,dx - \int_0^w f(-u + iy)e^{-iau - ay}\,dy \,.$$

Our choice of u and v ensures that $|v + iy|, |-u + iy| > R$ for all y in $[0, w]$, and $|x + iw| > R$ for all x in $[-u, v]$. Hence

$$\left| \int_0^w f(v + iy)e^{iav - ay}\,dy \right| \le \epsilon \left| \int_0^w e^{-ay}\,dy \right| = \frac{\epsilon}{a}\left(1 - e^{-aw}\right) \le \frac{\epsilon}{a}, \qquad (9.4)$$

and similarly

$$\left| \int_0^w f(-u + iy)e^{-iau - ay}\,dy \right| \le \epsilon \left| \int_0^w e^{-ay}\,dy \right| \le \frac{\epsilon}{a}, \qquad (9.5)$$

while (for sufficiently large w)

$$\left| \int_{-u}^{v} f(x+iw)e^{iax-aw}\, dx \right| \leq \epsilon \left| \int_{-u}^{v} e^{-aw}\, dx \right| = \epsilon w e^{-aw} \leq \epsilon. \qquad (9.6)$$

Hence, for sufficiently large u and v,

$$\left| \int_{-u}^{v} f(x)e^{iax}\, dx - 2\pi i \sum_{k=1}^{n} \mathrm{res}(g, c_k) \right|$$

can be made smaller than any positive number. Thus

$$\int_{-\infty}^{\infty} f(x)e^{iax}\, dx = 2\pi i \sum_{k=1}^{n} \mathrm{res}(g, c_k). \qquad (9.7)$$

$\square$

Remark 9.9

If, instead of the square, we had used the semicircle with centre 0 and radius R we would have concluded only that

$$(PV) \int_{-\infty}^{\infty} f(x)e^{iax}\, dx = 2\pi i \sum_{k=1}^{n} \mathrm{res}(g, c_k)$$

and it is possible for this to exist when the integral from $-\infty$ to ∞ does not. (See Section 1.7.)

Example 9.10

Evaluate

$$\int_{0}^{\infty} \frac{x \sin x\, dx}{x^2 + b^2}.$$

Solution

In Jordan's Lemma, let $f(z) = z/(z^2+b^2)$ and let $a = 1$. Thus $g(z) = ze^{iz}/(z^2+b^2)$. The only pole in the upper half-plane is at ib, and the residue is

$$\frac{ibe^{-b}}{2ib} = \frac{1}{2}e^{-b}.$$

Hence

$$\int_{-\infty}^{\infty} \frac{xe^{ix}\, dx}{x^2 + b^2} = \pi i e^{-b}. \qquad (9.8)$$

Equating imaginary parts gives

$$\int_{-\infty}^{\infty} \frac{x \sin x \, dx}{x^2 + b^2} = \pi e^{-b}.$$

The integrand is an even function, so

$$\int_0^{\infty} \frac{x \sin x \, dx}{x^2 + b^2} = \frac{\pi}{2} e^{-b}.$$

If we equate real parts in (9.4), we obtain

$$\int_{-\infty}^{\infty} \frac{x \cos x \, dx}{x^2 + b^2} = 0,$$

which we knew anyway, for the integrand is an odd function. □

Remark 9.11

Here we cannot deduce from the comparison test (Theorem 1.18) that the integral from 0 to ∞ converges.

Sometimes we need to make a small detour in our preferred contour so as to avoid, or sometimes to include, a pole. Before giving an example, we establish the following lemma:

Lemma 9.12

Suppose that f has a simple pole at c, with residue ρ, and let γ^* be a circular arc with radius r:

$$\gamma(\theta) = c + re^{i\theta} \quad (\alpha \le \theta \le \beta).$$

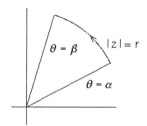

Then

$$\lim_{r \to 0} \int_{\gamma} f(z) \, dz = i\rho(\beta - \alpha).$$

Proof

In a suitable neighbourhood $N(c, \delta)$ we have a Laurent expansion

$$f(z) = \frac{\rho}{z - c} + \sum_{n=0}^{\infty} a_n (z - c)^n \,.$$

If we define $g(z)$ as $f(z) - \big(\rho/(z - c)\big)$, we see that g is bounded in $N(c, \delta)$. That is, there exists $M > 0$ such that $|g(z)| \leq M$, and so, if $0 < r < \delta$,

$$\left| \int_\gamma g(z)\, dz \right| \leq Mr(\beta - \alpha) \,.$$

Thus $\int_\gamma g(z)\, dz \to 0$ as $r \to 0$. Next, recall that

$$\int_\gamma \frac{\rho}{z - c}\, dz = \rho \int_\alpha^\beta \frac{rie^{i\theta}}{re^{i\theta}}\, d\theta = i\rho(\beta - \alpha) \,.$$

Given $\epsilon > 0$, it now follows that, for sufficiently small r,

$$\left| \int_\gamma f(z)\, dz - i\rho(\beta - \alpha) \right| = \left| \int_\gamma f(z)\, dz - \int_\gamma \frac{\rho}{z - c}\, dz \right|$$

$$= \left| \int_\gamma g(z)\, dz \right| \leq Mr(\beta - \alpha) < \epsilon \,.$$

Thus

$$\lim_{r \to 0} \int_\gamma f(z)\, dz = i\rho(\beta - \alpha)$$

as required.   □

Example 9.13

Show that

$$\int_0^\infty \frac{\sin x}{x}\, dx = \frac{\pi}{2} \,.$$

Solution

Consider e^{iz}/z, which has a simple pole at $z = 0$, with residue $e^{i0} = 1$. Except for the existence of this pole, the conditions of Jordan's Lemma are satisfied by the function $1/z$, and we modify the contour to a square with a small semicircular indentation σ to avoid 0.

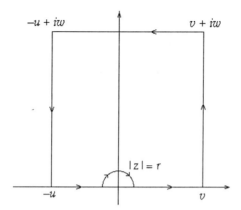

The inequalities (9.4), (9.5) and (9.6) all survive, and so, instead of (9.7), we have the modified conclusion that

$$\int_{-\infty}^{-r} \frac{e^{ix}}{x}\, dx - \int_{\sigma} \frac{e^{iz}}{z}\, dz + \int_{r}^{\infty} \frac{e^{ix}}{x}\, dx \; = \; 0\,. \tag{9.9}$$

By Lemma 9.12, as $r \to 0$,

$$\int_{\sigma} \frac{e^{iz}}{z}\, dz \; \to \; \pi i\,.$$

Hence, letting r tend to 0 and taking imaginary parts, we deduce that

$$\int_{-\infty}^{\infty} \frac{\sin x}{x}\, dx = \pi\,.$$

Since $\sin x/x$ is an even function, we obtain finally that

$$\int_{0}^{\infty} \frac{\sin x}{x}\, dx = \frac{\pi}{2}\,.$$

$\square$

Remark 9.14

Taking real parts in (9.9) gives

$$\int_{-\infty}^{-r} \frac{\cos x}{x}\, dx + \int_{r}^{\infty} \frac{\cos x}{x}\, dx \; \to \; 0$$

as $r \to 0$, which was obvious anyway, since $\cos x/x$ is an odd function. However, since $\cos x/x \sim 1/x$ as $x \to 0$, the integral $\int_{-\infty}^{\infty} (\cos x/x)\, dx$ does not exist. (See Section 1.7.)

EXERCISES

9.18. Evaluate
$$\int_{-\infty}^{\infty} \frac{\cos \pi x \, dx}{x^2 - 2x + 2}, \quad \int_{-\infty}^{\infty} \frac{\sin \pi x \, dx}{x^2 - 2x + 2}.$$

9.19. Show that
$$\int_{-\infty}^{\infty} \frac{x^3 \sin x}{(1 + x^2)^2} = \frac{\pi}{2e}.$$

9.20. Show that, for all real a,
$$\int_{-\infty}^{\infty} \frac{(x^2 - a^2) \sin x \, dx}{x(x^2 + a^2)} = \pi(2e^{-a} + 1).$$

9.21. Show that
$$\int_{0}^{\infty} \frac{\sin \pi x \, dx}{x(1 - x^2)} = \pi.$$

9.4 Real Integrals: Some Special Contours

An ingenious choice of contour can sometimes be used to compute a difficult integral.

Example 9.15

Evaluate
$$\int_{0}^{\infty} \frac{x}{1 + x^4} \, dx.$$

Solution

Here there is no point in finding the integral from $-\infty$ to ∞, since the integrand is an odd function, and this integral is trivially equal to 0. We consider $\int_{q(0,R)} [z/(1 + z^4)] \, dz$, where $q(0, R)$ is the quarter circle in the first quadrant.

The integrand has a simple pole at $e^{i\pi/4}$, with residue

$$\frac{e^{i\pi/4}}{4e^{3i\pi/4}} = \frac{1}{4}e^{-i\pi/2} = -\frac{i}{4},$$

and so

$$\int_0^R \frac{x}{1+x^4}\,dx + \int_0^{\pi/2} \frac{Re^{i\theta}}{1+R^4 e^{4i\theta}}\,iRe^{i\theta}\,d\theta - \int_0^R \frac{iy}{1+y^4}\,i\,dy = 2\pi i\left(-\frac{i}{4}\right) = \frac{\pi}{2}.$$

The middle term tends to 0 as $R \to \infty$, and the contribution of the section from iR to 0 is

$$-\int_0^R \frac{iy}{1+y^4}\,i\,dy = \int_0^R \frac{y}{1+y^4}\,dy\,,$$

the same as the contribution from the section from 0 to R. Letting $R \to \infty$, we deduce that

$$2\int_0^\infty \frac{x}{1+x^4}\,dx = \frac{\pi}{2}\,,$$

and so

$$\int_0^\infty \frac{x}{1+x^4}\,dx = \frac{\pi}{4}\,.$$

$\square$

Example 9.16

Evaluate

$$\int_{-\infty}^\infty \frac{e^{ax}}{1+e^x}\,dx \quad (0 < a < 1).$$

Solution

Here there is no problem over the convergence of the integral, since

$$\frac{e^{ax}}{1+e^x} \sim e^{-(1-a)x} \quad \text{as } x \to \infty,$$

and

$$\frac{e^{ax}}{1+e^x} \sim e^{ax} \quad \text{as } x \to -\infty.$$

We consider the complex function $f(z) = e^{az}/(1+e^z)$, and the rectangular contour with corners at $\pm R$, $\pm R + 2\pi i$.

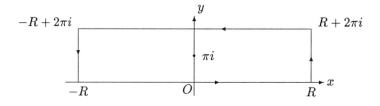

The function has a simple pole at πi, with residue $e^{a\pi i}/e^{\pi i} = e^{a\pi i}/(-1) = -e^{a\pi i}$. Hence

$$-2\pi i e^{a\pi i} = \int_{-R}^{R} \frac{e^{ax}}{1+e^x}\, dx + \int_{0}^{2\pi} \frac{e^{a(R+iy)}}{1+e^{R+iy}}\, i\, dy$$

$$- \int_{-R}^{R} \frac{e^{a(x+2\pi i)}}{1+e^{x+2\pi i}}\, dx - \int_{0}^{2\pi} \frac{e^{a(-R+iy)}}{1+e^{-R+iy}}\, i\, dy. \qquad (9.10)$$

Now, for sufficiently large R,

$$\left| \frac{e^{a(R+iy)}}{1+e^{R+iy}} \right| \leq \frac{e^{aR}}{e^R - 1} \leq 2e^{-(1-a)R}.$$

Hence

$$\left| \int_{0}^{2\pi} \frac{e^{a(R+iy)}}{1+e^{R+iy}}\, i\, dy \right| \leq 4\pi e^{-(1-a)R},$$

which tends to 0 as $R \to \infty$. Similarly, for sufficiently large R,

$$\left| \frac{e^{a(-R+iy)}}{1+e^{-R+iy}} \right| \leq \frac{e^{-aR}}{1-e^{-R}} \leq 2e^{-aR}.$$

Hence

$$\left| \int_{0}^{2\pi} \frac{e^{a(-R+iy)}}{1+e^{-R+iy}}\, i\, dy \right| \leq 4\pi e^{-aR},$$

which again tends to 0 as $R \to \infty$. Also,

$$\int_{-R}^{R} \frac{e^{a(x+2\pi i)}}{1+e^{x+2\pi i}}\, dx = e^{2a\pi i} \int_{-R}^{R} \frac{e^{ax}}{1+e^x}\, dx.$$

Hence, letting $R \to \infty$, we see from (9.10) that

$$(1 - e^{2a\pi i}) \int_{-\infty}^{\infty} \frac{e^{ax}}{1+e^x}\, dx = -2\pi i e^{a\pi i}.$$

Hence

$$\int_{-\infty}^{\infty} \frac{e^{ax}}{1+e^x}\, dx = \frac{2\pi i}{e^{a\pi i} - e^{-a\pi i}} = \frac{\pi}{\sin a\pi}. \qquad (9.11)$$

$\square$

Remark 9.17

If we substitute $x = \log t$ in this integral, we obtain

$$\int_{0}^{\infty} \frac{t^{a-1}}{1+t}\, dt,$$

and if you have come across real beta- and gamma-functions (see [13]) you may recognise this last integral as $B(a, 1-a) = \Gamma(a)\Gamma(1-a)$. (The further substitution $t = u/(1-u)$ changes the integral to $\int_0^1 u^{a-1}(1-u)^{-a}\,du$, which relates to the usual definition $B(m,n) = \int_0^1 x^{m-1}(1-x)^{n-1}\,dx$ of the B-function.) Formula (9.11) thus gives us the identity

$$\Gamma(a)\Gamma(1-a) = \frac{\pi}{\sin a\pi} \quad (0 < a < 1). \tag{9.12}$$

Following on from this remark, we see that from (9.12) it follows, by putting $a = \frac{1}{2}$, that

$$\Gamma\left(\tfrac{1}{2}\right) = \sqrt{\pi}. \tag{9.13}$$

From [13] we have the definition

$$\Gamma(n) = \int_0^\infty t^{n-1}e^{-t}\,dt = 2\int_0^\infty x^{2n-1}e^{-x^2}\,dx.$$

Thus

$$\int_0^\infty e^{-x^2}\,dx = \tfrac{1}{2}\Gamma\left(\tfrac{1}{2}\right) = \tfrac{1}{2}\sqrt{\pi}. \tag{9.14}$$

Since e^{-x^2} is an even function, we can deduce that

$$\int_{-\infty}^\infty e^{-x^2}\,dx = \sqrt{\pi}. \tag{9.15}$$

These integrals are important in probability theory, since e^{-x^2} occurs as the probability density function of the normal distribution. See, for example, [6].

Here I have been breaking the rules by referring to aspects of real analysis not mentioned in Chapter 1. It is in fact possible to obtain the integral (9.15) by contour integration, and the beautifully ingenious proof is a rare delight:

Theorem 9.18

$$\int_{-\infty}^\infty e^{-x^2}\,dx = \sqrt{\pi}.$$

Proof

Let $f(z) = e^{i\pi z^2}$, and let $g(z) = f(z)/\sin \pi z$. Let $c = e^{i\pi/4} = (1/\sqrt{2})(1+i)$. Note that $c^2 = i$. We integrate $g(z)$ round the parallelogram with vertices

$\pm Rc \pm \frac{1}{2}$.

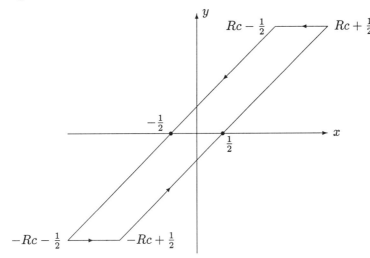

Since the zeros of $\sin \pi z$ are at $0, \pm 1, \pm 2, \ldots$, the only pole of g within the contour is at 0, where the residue is $1/\pi$. Hence the integral round the contour has the value $2i$. The sloping sides can be parametrised (respectively) by

$$\sigma_1(t) = ct + \tfrac{1}{2}, \quad \sigma_2(t) = ct - \tfrac{1}{2} \quad (-R \le t \le R).$$

Now,

$$\begin{aligned}
f(ct + \tfrac{1}{2}) &= \exp\left[i\pi(ct + \tfrac{1}{2})^2\right] \\
&= \exp\left[i\pi(it^2 + ct + \tfrac{1}{4})\right] \quad \text{(since } c^2 = i) \\
&= \exp[-\pi t^2 + \pi i ct + (i\pi/4)],
\end{aligned}$$

and so, since $\sin \pi(ct + \tfrac{1}{2}) = \cos \pi ct$,

$$g(ct + \tfrac{1}{2}) = \exp[-\pi t^2 + \pi i ct + (i\pi/4)]/\cos \pi ct.$$

Similarly, since $\sin \pi(ct - \tfrac{1}{2}) = -\cos \pi ct$,

$$g(ct - \tfrac{1}{2}) = - \exp[-\pi t^2 - \pi i ct + (i\pi/4)]/\cos \pi ct.$$

Hence the combined contribution to the integral of g of the sections of the contour from $-Rc + \tfrac{1}{2}$ to $Rc + \tfrac{1}{2}$ and from $Rc - \tfrac{1}{2}$ to $-Rc - \tfrac{1}{2}$ is

$$\int_{-R}^{R} \frac{\exp(-\pi t^2) \exp(i\pi/4)}{\cos \pi ct} \left(e^{i\pi ct} + e^{-i\pi ct}\right) c \, dt = \int_{-R}^{R} \frac{2c^2 \exp(-\pi t^2) \cos \pi ct}{\cos \pi ct} \, dt$$

$$= 2i \int_{-R}^{R} e^{-\pi t^2} \, dt.$$

As for the contribution of the horizontal section from $Rc - \frac{1}{2}$ to $Rc + \frac{1}{2}$, for all u in $[-\frac{1}{2}, \frac{1}{2}]$,

$$\begin{aligned} |\exp[i\pi(Rc + u)^2]| &= |\exp[i\pi(R^2 i + \sqrt{2}Ru(1 + i) + u^2)]| \\ &= \exp[-\pi(R^2 + \sqrt{2}Ru)] \le \exp[-\pi(R^2 - (R/\sqrt{2}))] \,, \end{aligned}$$

and, from Exercise 4.13, we have that

$$|\sin[\pi(Rc + u)]| = |\sin[(\pi/\sqrt{2})(R + iR + \sqrt{2}\,u)]| \ge \sinh(\pi R/\sqrt{2}) \,.$$

Hence

$$\left| \int_{-1/2}^{1/2} g(Rc + u)\, du \right| \le \frac{\exp[-\pi(R^2 - (R/\sqrt{2}))]}{\sinh(\pi R/\sqrt{2})} \,,$$

and this certainly tends to 0 as $R \to \infty$. A similar argument establishes that, as $R \to \infty$,

$$\int_{-1/2}^{1/2} g(-Rc + u)\, du \to 0 \,.$$

Letting $R \to \infty$, we have

$$2i \int_{-\infty}^{\infty} e^{-\pi t^2}\, dt = 2i$$

and so

$$\int_{-\infty}^{\infty} e^{-\pi t^2}\, dt = 1 \,.$$

The proof is completed by substituting $x = t/\sqrt{\pi}$ in this last integral.   □

Example 9.19

Show that

$$\int_0^{\infty} e^{-x^2} \cos 2ax\, dx = \tfrac{1}{2}\sqrt{\pi}\, e^{-a^2} \,.$$

Solution

It is sufficient to consider $a > 0$, since cos is an even function. We consider the function e^{-z^2} and integrate round the rectangle with vertices 0, R, $R + ia$, ia.

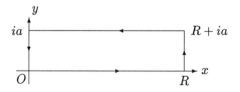

The function e^{-z^2} has no poles, and so

$$0 = \int_0^R e^{-x^2}\,dx + \int_0^a e^{-(R+iy)^2}i\,dy - \int_0^R e^{-(x+ia)^2}\,dx - \int_0^a e^{y^2}i\,dy\,. \quad (9.16)$$

Now,

$$\int_0^R e^{-(x+ia)^2}\,dx = e^{a^2}\int_0^R e^{-x^2}e^{-2iax}\,dx$$

$$= e^{a^2}\int_0^R e^{-x^2}(\cos 2ax - i\sin 2ax)\,dx\,.$$

Also,

$$\left|\int_0^a e^{-(R+iy)^2}i\,dy\right| = e^{-R^2}\left|\int_0^a e^{-2iRy}e^{y^2}i\,dy\right| \le e^{-R^2}ae^{a^2}\,,$$

and this tends to 0 as $R \to \infty$. Hence, letting $R \to \infty$ in (9.16), taking real parts, and noting that the final term of (9.16) is pure imaginary, we find that

$$0 = \int_0^\infty e^{-x^2}\,dx - e^{a^2}\int_0^\infty e^{-x^2}\cos 2ax\,dx\,.$$

Hence, from (9.14),

$$\int_0^\infty e^{-x^2}\cos 2ax\,dx = \tfrac{1}{2}\sqrt{\pi}e^{-a^2}\,.$$

$\square$

The solution of the next example requires a simple result in real analysis:

Lemma 9.20

If $0 < \theta \le \pi/2$, then

$$\frac{2}{\pi} \le \frac{\sin\theta}{\theta} \le 1\,.$$

Proof

Since $\lim_{\theta\to 0}(\sin\theta/\theta) = 1$, and since $[\sin(\pi/2)]/(\pi/2) = 2/\pi$, it is enough to show that $\sin\theta/\theta$ decreases throughout the interval $(0, \pi/2]$. The derivative is

$$\frac{\theta\cos\theta - \sin\theta}{\theta^2}\,,$$

and so the problem reduces to showing that $F(\theta) = \theta\cos\theta - \sin\theta \le 0$ throughout $[0, \pi/2]$. This is clear, since $F(0) = 0$ and $F'(\theta) = -\theta\cos\theta$ is certainly non-positive in $[0, \pi/2]$. $\square$

Example 9.21

Evaluate

$$\int_0^\infty \cos(x^2)\, dx \, .$$

Solution

Here again we devise a contour where the sections that do not tend to 0 both contribute to the answer. We consider $\int_\gamma \exp(iz^2)\, dz$ where γ is as shown:

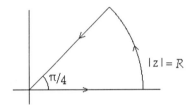

The function is holomorphic throughout, and so

$$0 = \int_0^R \exp(ix^2)\, dx + \int_0^{\pi/4} \exp(iR^2 e^{2i\theta})iRe^{i\theta}\, d\theta - \int_0^R \exp(it^2 e^{i\pi/2})e^{i\pi/4}\, dt$$
$$= I_1 + I_2 - I_3 \quad \text{(say)}\,.$$

Now,

$$|\exp(iR^2 e^{2i\theta})iRe^{i\theta}| = |R\exp\left(R^2(i\cos 2\theta - \sin 2\theta)\right)| = R\exp(-R^2 \sin 2\theta)$$
$$\leq R\exp(-4R^2\theta/\pi) \quad \text{(by Lemma 9.20)}\,.$$

Hence

$$|I_2| \leq R\int_0^{\pi/4} \exp(-4R^2\theta/\pi)\, d\theta = \frac{\pi}{4R}[1 - \exp(-R^2)]\,,$$

which tends to 0 as $R \to \infty$. Hence, letting $R \to \infty$ and using (9.14), we deduce that

$$\int_0^\infty e^{ix^2}\, dx = \lim_{R\to\infty} I_3 = \frac{1+i}{\sqrt{2}}\int_0^\infty e^{-t^2}\, dt = \frac{(1+i)\sqrt{\pi}}{2\sqrt{2}}\,.$$

Taking real and imaginary parts, we see that

$$\int_0^\infty \cos(x^2)\, dx = \int_0^\infty \sin(x^2)\, dx = \sqrt{\pi/8}.$$

$\square$

EXERCISES

9.22. By considering the integral of $e^{az}/\cosh z$ round the rectangular contour

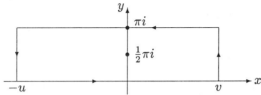

show that

$$\int_{-\infty}^{\infty} \frac{e^{ax}}{\cosh x}\, dx = \frac{\pi}{\cos(\pi a/2)} \quad (-1 < a < 1).$$

9.23. By integrating $f(z) = z^{4n+3}e^{-z}$ round the contour γ shown,

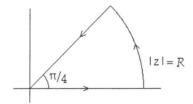

prove that

$$\int_0^{\infty} x^{4n+3}e^{-x}\cos x\, dx = (-1)^{n+1}\frac{(4n+3)!}{2^{2n+2}}.$$

You may assume that $\int_0^{\infty} x^n e^{-x}\, dx = n!$.

9.24. Show that, for all $a, b > 0$,

$$\int_{-\infty}^{\infty} \frac{\cos ax - \cos bx}{x^2}\, dx = \pi(b - a).$$

Deduce that

$$\int_0^{\infty} \left(\frac{\sin x}{x}\right)^2 dx = \frac{\pi}{2}.$$

9.25. By considering the integral of $z/(a - e^{-iz})$ round the rectangle with vertices $\pm\pi$, $\pm\pi + iR$, show that

$$\int_0^{\pi} \frac{x \sin x}{1 - 2a\cos x + a^2}\, dx = \frac{\pi}{a}\log(1 + a) \quad (0 < a < 1).$$

9.5 Infinite Series

One of the more unexpected applications of the residue theorem is to the summation of infinite series. The key to the technique is the observation that $\cot \pi z$ and $\operatorname{cosec} \pi z$ both have simple poles at each integer n. These in fact are the only poles, since $\sin \pi z = 0$ if and only if z is real and an integer.

We begin with an example.

Example 9.22

Show that

$$\sum_{n=1}^{\infty} \frac{1}{n^2 + a^2} = \frac{1}{2a} \coth \pi a - \frac{1}{2a^2} \quad (0 < a < 1).$$

Solution

Let

$$g(z) = \frac{\pi \cot \pi z}{z^2 + a^2} \quad (0 < a < 1),$$

and consider the square contour σ_n with vertices at $(n + \frac{1}{2})(\pm 1 \pm i)$.

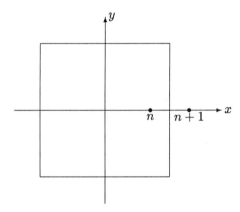

Within the contour, the function g has simple poles at ai, $-ai$ and $\pm 1, \pm 2, \ldots \pm n$. The residue of the simple pole (see Exercise 8.12) at the non-zero integer k is $1/(k^2 + a^2)$. The residue at ai is

$$\lim_{z \to ai} \frac{\pi \cot \pi z}{z + ai} = \frac{1}{2ai} \pi \cot i \pi a = -\frac{1}{2a} \pi \coth \pi a,$$

and a similar calculation shows that this is also the residue at $-ai$. Hence

$$\int_{\sigma_n} g(z)\, dz = 2\pi i \left(\sum_{k=-n}^{n} \frac{1}{n^2 + a^2} - \frac{1}{a} \pi \coth \pi a \right). \tag{9.17}$$

We now examine what happens as $n \to \infty$, and for this it is useful to prove a lemma, only half of which we need immediately.

Lemma 9.23

Let n be a positive integer and let σ_n be the square with corners at $(\pm 1 \pm i)(n + \frac{1}{2})$. Then there exist constants A, B such that, for all n and for all z on the square σ_n,

$$|\cot \pi z| \leq A \quad |\operatorname{cosec} \pi z| \leq B.$$

Proof

On the horizontal sides, where $z = x \pm i(n + \frac{1}{2})$,

$$\begin{aligned}
|\cot \pi z| &= \left| \frac{e^{i\pi[x \pm i(n+\frac{1}{2})]} + e^{-i\pi[x \pm i(n+\frac{1}{2})]}}{e^{i\pi[x \pm i(n+\frac{1}{2})]} - e^{-i\pi[x \pm i(n+\frac{1}{2})]}} \right| \\
&\leq \frac{e^{\pi(n+\frac{1}{2})} + e^{-\pi(n+\frac{1}{2})}}{e^{\pi(n+\frac{1}{2})} - e^{-\pi(n+\frac{1}{2})}} \\
&= \coth(n + \tfrac{1}{2})\pi \\
&\leq \coth(\tfrac{3}{2}\pi),
\end{aligned}$$

since coth is a decreasing function in the interval $(0, \infty)$. On the vertical sides, where $z = \pm(n + \frac{1}{2}) + iy$, using the properties that $\cot(z + \frac{1}{2}\pi) = -\tan z$ and $\cot(z + n\pi) = \cot z$, we see that

$$|\cot \pi z| = |-\tan \pi i y| = |\tanh \pi y| \leq 1.$$

(See Exercise 4.7.)

As for $\operatorname{cosec} \pi z$, on the vertical sides we make use of the equalities $\operatorname{cosec}(z + \frac{1}{2}\pi) = \sec z$ and $\operatorname{cosec}(z + n\pi) = (-1)^n \operatorname{cosec} z$ to show that

$$|\operatorname{cosec} \pi[\pm(n + \tfrac{1}{2}) + iy]| = |\sec iy| = |\operatorname{sech} y| \leq 1.$$

(See Exercise 4.7.) On the horizontal sides, where $z = x \pm i(n + \frac{1}{2})$,

$$|\operatorname{cosec} \pi z| = \left| \frac{1}{e^{i\pi[x \pm i(n+\frac{1}{2})]} - e^{-i\pi[x \pm i(n+\frac{1}{2})]}} \right| \leq \frac{1}{e^{\pi(n+\frac{1}{2})} - e^{-\pi(n+\frac{1}{2})}}$$

$$= \operatorname{cosech}(n + \tfrac{1}{2})\pi \leq \operatorname{cosech}(\tfrac{3}{2}\pi),$$

since cosech is a decreasing function in the interval $(0, \infty)$.   □

It now follows, since the contour σ_n is of total length $4(2n+1)$, and since $\min\{|z| : z \in \sigma_n\} = n + \frac{1}{2}$, that

$$\left| \int_{\sigma_n} g(z)\, dz \right| \leq 4(2n+1)\pi \sup_{z \in \sigma_n} \left| \frac{\cot \pi z}{z^2 + a^2} \right| \leq 4(2n+1) \frac{\pi A}{(n+\frac{1}{2})^2 - a^2},$$

and this tends to 0 as $n \to \infty$. Hence, letting $n \to \infty$ in (9.17), we find that

$$\sum_{n=-\infty}^{\infty} \frac{1}{n^2 + a^2} = \frac{\pi}{a} \coth \pi a. \qquad (9.18)$$

That is,

$$\frac{1}{a^2} + 2 \sum_{n=1}^{\infty} \frac{1}{n^2 + a^2} = \frac{\pi}{a} \coth \pi a,$$

and so

$$\sum_{n=1}^{\infty} \frac{1}{n^2 + a^2} = \frac{\pi}{2a} \coth \pi a - \frac{1}{2a^2}. \qquad (9.19)$$

This example is an instance of a general result:

Theorem 9.24

Let f be a function that is differentiable on $\mathbb{C}$ except for finitely many poles $c_1, c_2, \ldots, c_m$, none of which is a real integer. Suppose also that there exist $K, R > 0$ such that $|z^2 f(z)| \leq K$ whenever $|z| > R$. Let

$$g(z) = \pi f(z) \cot \pi z \qquad h(z) = \pi f(z) \operatorname{cosec} \pi z.$$

Then

$$\sum_{n=-\infty}^{\infty} f(n) = - \sum_{k=1}^{m} \operatorname{res}(g, c_k), \qquad (9.20)$$

$$\sum_{n=-\infty}^{\infty} (-1)^n f(n) = - \sum_{k=1}^{m} \operatorname{res}(h, c_k). \qquad (9.21)$$

Proof

Certainly the series $\sum f(n)$ and $\sum (-1)^n f(n)$ are (absolutely) convergent, since $|f(n)| < K/n^2$ whenever $|n| > R$. For both $\pi \cot \pi z$ and $\pi \operatorname{cosec} \pi z$ the set of poles is precisely the set of real integers, and so for both g and h the set of poles is $\{c_1, c_2, \ldots, c_m\} \cup \mathbb{Z}$. The residue of g at the integer n is $f(n)$, and the residue of h at n is $(-1)^n f(n)$. (See Exercise 8.12.) We use the square contour σ_n of Example 9.22 . Then, by Lemma 9.23, if n is large enough,

$$\left| \int_{\sigma_n} g(z)\, dz \right| \leq 4(2n+1) \sup_{z \in \sigma_n} |\pi f(z) \cot \pi z| \leq 4(2n+1) \frac{KA}{n^2},$$

and

$$\left| \int_{\sigma_n} h(z)\, dz \right| \leq 4(2n+1) \sup_{z \in \sigma_n} |\pi f(z) \operatorname{cosec} \pi z| \leq 4(2n+1) \frac{KB}{n^2},$$

and these both tend to 0 as $n \to \infty$. Thus, letting $n \to \infty$, we obtain the equalities (9.20) and (9.21). $\qquad\square$

Example 9.25

Show that

$$\sum_{n=0}^{\infty} \frac{(-1)^n}{(2n+1)^3} = \frac{\pi^3}{32}.$$

Solution

The function

$$h(z) = \frac{\pi}{(2z+1)^3 \sin \pi z}$$

has a pole of order 3 at $-\frac{1}{2}$. The residue is $\frac{1}{2} q''(-\frac{1}{2})$, where

$$q(z) = (z + \tfrac{1}{2})^3 h(z) = \frac{\pi}{8} \frac{1}{\sin \pi z}.$$

Since

$$q'(z) = -\frac{\pi^2}{8} \frac{\cos \pi z}{\sin^2 \pi z}, \quad q''(z) = -\frac{\pi^2}{8} \frac{-\sin^2 \pi z\, \pi \sin \pi z - 2 \sin \pi z\, \pi \cos \pi z}{\sin^4 \pi z}$$

we see that $\operatorname{res}(h, -\frac{1}{2}) = -\pi^3/16$. Hence, by Theorem 9.24,

$$\frac{\pi^3}{16} = \cdots - \frac{1}{(-5)^3} + \frac{1}{(-3)^3} - \frac{1}{(-1)^3} + \frac{1}{1^3} - \frac{1}{3^3} + \frac{1}{5^3} - \cdots$$

$$= 2\left(\frac{1}{1^3} - \frac{1}{3^3} + \frac{1}{5^3} - \cdots \right),$$

and so, as required,

$$\sum_{n=0}^{\infty} \frac{(-1)^n}{(2n+1)^3} = \frac{\pi^3}{32}.$$

$\qquad\square$

The requirement in Theorem 9.24 that f have no poles at real integers can be relaxed, simply by treating the "offending" integer on its own.

Example 9.26

Show that

$$\sum_{n=1}^{\infty} \frac{1}{n^2} = \frac{\pi^2}{6} .$$

Solution

Here the function

$$g(z) = \frac{\pi \cos \pi z}{z^2 \sin \pi z}$$

has simple poles at $\pm 1, \pm 2, \ldots$, and a triple pole at 0. The residue at each non-zero integer n is $1/n^2$, and at 0 (see Exercise 8.11) the residue is $-\pi^2/3$. Hence

$$\sum_{n \in \mathbb{Z} \setminus \{0\}} \frac{1}{n^2} = \frac{\pi^2}{3} ,$$

and so

$$\sum_{n=1}^{\infty} \frac{1}{n^2} = \frac{\pi^2}{6} .$$

$\square$

Remark 9.27

Alternatively one could deduce the value of $\sum_{n=1}^{\infty}(1/n^2)$ from (9.19) by letting $a \to 0$. The limiting process is legitimate since the series $\sum_{n=1}^{\infty}[1/(n^2 + a^2)]$ is uniformly convergent for a in $[-1, 1]$. The actual calculation is a pleasant exercise on L'Hôpital's Rule.

EXERCISES

9.26. Sum the series

$$\sum_{n=1}^{\infty} \frac{(-1)^n}{n^2 + a^2} .$$

9.27. Show that, if a is not an integer, then

$$\sum_{n=-\infty}^{\infty} \frac{1}{(n + a)^2} = \pi^2 \operatorname{cosec}^2 \pi a .$$

Deduce the values of

$$\sum_{-\infty}^{\infty} \frac{1}{(2n + 1)^2} , \quad \sum_{-\infty}^{\infty} \frac{1}{(3n + 1)^2} , \quad \sum_{-\infty}^{\infty} \frac{1}{(4n + 1)^2} .$$

9.28. Show that, if a is not an integer,

$$\sum_{n=1}^{\infty} \frac{1}{n^4 - a^4} = \frac{1}{2a^4} - \frac{\pi}{4a^3}\left(\cot \pi a + \coth \pi a\right).$$

10
Further Topics

10.1 Integration of f'/f; Rouché's Theorem

In this section we examine an integral that in effect counts the number of poles and zeros of a meromorphic function f. Recall that, if f has Laurent series $\sum_{n=-\infty}^{\infty} a_n(z-c)^n$ at c, then $\operatorname{ord}(f,c) = \min\{n : a_n \neq 0\}$. If $\operatorname{ord}(f,c) = m > 0$ then $f(c) = 0$, and we say that c is a **zero of order** m of the function f. If $\operatorname{ord}(f,c) = -m < 0$, then c is a **pole of order** m.

Theorem 10.1

Let γ be a contour, let f be meromorphic in a domain that contains $\mathrm{I}(\gamma) \cup \gamma^*$, and suppose that $Q = \{q \in \mathrm{I}(\gamma) : \operatorname{ord}(f,q) \neq 0\}$ is finite. Then

$$\frac{1}{2\pi i} \int_\gamma \frac{f'(z)}{f(z)}\, dz = \sum_{q \in Q} \operatorname{ord}(f,q)\,.$$

Proof

The function f'/f is differentiable in $\mathrm{I}(\gamma) \setminus Q$. Let $q \in Q$ and suppose that $\operatorname{ord}(f,q) = m \neq 0$. Then $f(z) = (z-q)^m g(z)$, where g is holomorphic and non-zero at q. Hence

$$f'(z) = m(z-q)^{m-1}g(z) + (z-q)^m g'(z)\,,$$

and so

$$\frac{f'(z)}{f(z)} = \frac{m}{z-q} + \frac{g'(z)}{g(z)}\,.$$

Since g'/g is holomorphic at q, we deduce that

$$\operatorname{res}(f'/f, q) = m = \operatorname{ord}(f, q)\,.$$

The result now follows from the Residue Theorem.   □

Remark 10.2

Informally, this result says that the integral of f'/f is $2\pi i$ times the number of zeros minus the number of poles, where each is counted according to its order. Thus, for example, if $\gamma = \kappa(0, 6)$ and

$$f(z) = \frac{(z-1)(z-2)^3}{(z-3)^2(z-4)^2(z-5)^2}\,,$$

then

$$\int_\gamma \frac{f'(z)}{f(z)}\, dz = 2\pi i[(1+3)-(2+2+2)] = -4\pi i\,.$$

The fact, on the face of it surprising, that $(1/2\pi i)\int_\gamma(f'/f)$ is an integer has the following interesting consequence, as observed by Rouché[1] :

Theorem 10.3 (Rouché's Theorem)

Let γ be a contour, and let f, g be functions defined in an open domain D containing $\mathrm{I}(\gamma) \cup \gamma^*$. Suppose also that:

(i) f and g are differentiable in D, except for a finite number of poles, none lying on γ^*;

(ii) f and $f + g$ have at most finitely many zeros in D;

(iii) $|g(z)| < |f(z)|$ for all z in γ^*.

Then

$$\int_\gamma \frac{f'(z)+g'(z)}{f(z)+g(z)}\, dz = \int_\gamma \frac{f(z)}{g(z)}\, dz\,.$$

---

[1] Eugène Rouché, 1832–1910.

Proof

The basic idea of the proof is that a small change in the function f should bring about a small change in $(1/2\pi i)\int_\gamma (f'/f)$. But integers are not amenable to small changes, and so, if the change in f is sufficiently small, as measured by Condition (iii) in the statement of the theorem, then the change must be zero.

By Corollary 5.4,

$$\inf\{|f(z)| - |g(z)| : z \in \gamma^*\} = \delta > 0. \tag{10.1}$$

Let $0 \le t \le 1$, and let

$$J(t) = \frac{1}{2\pi i}\int_\gamma \frac{f' + tg'}{f + tg}.$$

If $f(c) + tg(c)$ were 0 for some c on γ^*, then we would have $|g(c)| = |f(c)|/t \ge |f(c)|$, in contradiction to (iii) above. Hence $f + tg$ has no zeros on γ^*.

We show that $J(t)$ is continuous. Let $0 \le t < u \le 1$. Then

$$\left|\frac{(f' + ug')(z)}{(f + ug)(z)} - \frac{(f' + tg')(z)}{(f + tg)(z)}\right| = \left|\frac{(u - t)(fg' - f'g)(z)}{(f + ug)(z)(f + tg)(z)}\right|.$$

Now, $fg' - f'g$, being continuous on γ^*, is bounded, by Theorem 5.3: say $|(fg' - f'g)(z)| \le M$ for all z on γ^*. Also, from (10.1), $|(f + ug)(z)(f + tg)(z)| \ge \delta^2$. Hence

$$\left|\frac{(f' + ug')(z)}{(f + ug)(z)} - \frac{(f' + tg')(z)}{(f + tg)(z)}\right| \le \frac{M|u - t|}{\delta^2},$$

which can be made less than any positive ϵ by choosing $|u - t|$ sufficiently small.

Let Λ denote the length of the contour γ. Then

$$|J(u) - J(t)| \le \frac{\Lambda}{2\pi}\sup\left|\frac{f' + ug'}{f + ug} - \frac{f' + tg'}{f + tg}\right| \le \frac{\Lambda M}{2\pi\delta^2}|u - t| = K|u - t|,$$

where K is a constant independent of u and t. If $|u - t| < 1/K$, then $|J(u) - J(t)| < 1$ and so, since $J(u)$ and $J(t)$ are both integers, $J(u) = J(t)$. Hence, if $n > K$,

$$J(0) = J(1/n),\ J(1/n) = J(2/n),\ \ldots, J\big((n-1)/n\big) = J(1),$$

and so $J(0) = J(1)$. $\qquad\qquad\square$

The Fundamental Theorem of Algebra (see Theorem 7.11) is an immediate corollary. Let

$$p(z) = a_n z^n + a_{n-1}z^{n-1} + \cdots + a_1 z + a_0 = a_n z^n + q(z)$$

be a polynomial of degree n. On a circle $\kappa(n, R)$ with sufficiently large radius, $|q(z)| \leq |a_n z^n|$. Neither function has any poles, and so the number of zeros (counted according to order) of $a_n z^n + q(z)$ is the same as that of $a_n z^n$. The latter function has a single zero of order n, and the theorem follows.

Another interpretation of $\int_\gamma (f'/f)$ is useful. In Section 7.3 we encountered the function $\log z$, defined within a contour γ such that $0 \notin I(\gamma) \cup \gamma^*$. Certainly $(\log)'\big(f(z)\big) = f'(z)/f(z)$, and so, recalling Theorems 5.18 and 5.19, we can interpret $\int_\gamma (f'/f)$ as measuring the change in $\log\big(f(z)\big)$ as z moves round the contour γ. If f has no zeros or poles within γ, then f'/f is holomorphic, and $\int_\gamma (f'/f) = 0$. In this case there are no branch points of $\log\big(f(z)\big)$ within γ, and the function returns to its original value. Since

$$\log\big(f(z)\big) = \log|f(z)| + i \arg\big(f(z)\big),$$

we can equally well interpret the integral $\int_\gamma$ as measuring the change in $\arg\big(f(z)\big)$ as z moves round γ. This observation is frequently referred to as the **Principle of the Argument**.

A very simple example illustrates the point. If $f(z) = z^3$, then $f'(z)/f(z) = 3/z$, and the integral of $3/z$ round the unit circle is $6\pi i$. (This is in accord with Theorem 10.1, since f has a triple pole at 0.) The change in $\arg(z^3)$ is 6π, since z^3 goes round the circle three times. This prompts a definition:

$$\Delta_\gamma(\arg f) = \frac{1}{i} \int_\gamma \frac{f'}{f}. \tag{10.2}$$

Thus, in the notation of Theorem 10.1,

$$\Delta_\gamma(\arg f) = 2\pi \sum_{z \in Q} \operatorname{ord}(f, z). \tag{10.3}$$

We easily see that Δ_γ behaves logarithmically:

$$\Delta_\gamma\big(\arg(fg)\big) = \Delta_\gamma(\arg f) + \Delta_\gamma(\arg g);$$

for

$$\Delta_\gamma\big(\arg(fg)\big) = \frac{1}{i} \int_\gamma \frac{(fg)'}{fg} = \frac{1}{i} \int_\gamma \frac{f'g + fg'}{fg} = \frac{1}{i} \int_\gamma \frac{f'}{f} + \frac{1}{i} \int_\gamma \frac{g'}{g}$$
$$= \Delta_\gamma(\arg f) + \Delta_\gamma(\arg g).$$

In a similar manner one can show that

$$\Delta_\gamma\big(\arg(1/f)\big) = -\Delta_\gamma(\arg f).$$

The argument changes continuously as z moves round the contour, and if we divide the contour γ into pieces γ_1 and γ_2,

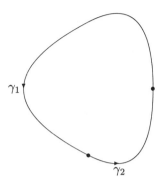

then

$$\Delta_\gamma(\arg f) = \Delta_{\gamma_1}(\arg f) + \Delta_{\gamma_2}(\arg f)\,.$$

The Fundamental Theorem of Algebra establishes that every polynomial of degree n has n roots, but gives no indication of where in the complex plane these roots might be. We can sometimes use the principle of the argument to be more specific over the location of roots.

Example 10.4

Show that the function $f(z) = z^4 + z^3 + 1$ has one zero in the first quadrant.

Solution

We can use elementary calculus to establish that the equation $x^4 + x^3 + 1 = 0$ has no real roots, for $x^4 + x^3 + 1$ has a minimum value of $229/256$ when $x = -3/4$. There are no zeros on the y-axis either, since $f(iy) = (y^4 + 1) - iy^3$, and this cannot be zero. Let R be real and positive, and consider a contour γ consisting of the line segment γ_1 from 0 to R, the circular arc γ_2 from R to iR and the line segment γ_3 from iR to 0. It is clear that $\arg f(z)$ has the constant value 0 throughout γ_1, and so $\Delta_{\gamma_1}(\arg f) = 0$. On γ_3 the argument is $\tan^{-1}\big(-y^3/(y^4 + 1)\big)$. This has the value 0 when $y = 0$, and its value at $y = R$ tends to 0 as $R \to \infty$. Thus $\Delta_{\gamma_3}(\arg f) \to 0$ as $R \to \infty$ Coming finally to γ_2, we use Rouché's Theorem to observe that, for sufficiently large R,

$$\Delta_{\gamma_2}(\arg f) = \Delta_{\gamma_2}(\arg z^4) \to 2\pi \ \text{ as } R \to \infty\,.$$

Thus $\Delta_\gamma(\arg f) = 2\pi$ for sufficiently large R, and so there is one root of $f(z) = 0$ in the first quadrant. ☐

EXERCISES

10.1. Show that the equation $z^8 + 3z^3 + 7z + 5 = 0$ has two roots in the first quadrant. Are the roots distinct?

10.2. Let $a > e$. Show that the equation $e^z = az^n$ has n roots inside the circle $\kappa(0, 1)$.

10.3. Show that the polynomial $z^5 + 15z + 1$ has precisely four zeros in the annular region $\{z : 3/2 < |z| < 2\}$.

10.4. Show that the equation $z^5 + 7z + 12 = 0$ has one root on the negative real axis. Show also that there is in addition one root in each quadrant, and that all the roots are in the annulus $\{z : 1 < |z| < 2\}$.

10.5. Let $n \geq 3$. Show that the polynomial $z^n + nz - 1$ has n zeros in $N(0, R)$, where

$$R = 1 + \left(\frac{2}{n-1}\right)^{1/2} .$$

10.2 The Open Mapping Theorem

In this section we explore some further properties of holomorphic functions. The first observation is that, unless the holomorphic function f is identically zero, the zeros of f are **isolated**. Precisely, we have:

Theorem 10.5

Let f be holomorphic in an open set U and let $c \in U$ be such that $f(c) = 0$. Then, unless f is the zero function, there exists $\delta > 0$ such that $f(z)$ is non-zero for all z in the punctured disc $D'(c, \delta)$.

Proof

Since f is holomorphic in an open set containing c, it has a Taylor series: within $N(c, r)$,

$$f(z) = \sum_{n=0}^{\infty} a_n (z - c)^n .$$

If f is the zero function then all the coefficients a_n are zero. Otherwise there exists $m > 0$ such that $a_m \neq 0$ and

$$f(z) = a_m (z - c)^m + a_{m+1}(z - c)^{m+1} + \cdots .$$

Let

$$g(z) = (z - c)^{-m} f(z) = a_m + a_{m+1}(z - c) + \cdots.$$

Then $g(c) \neq 0$ and so, since g is continuous at c, there exists $\delta > 0$ such that $g(z) \neq 0$ for all z in $N(c, \delta)$. Since $f(z) = (z - c)^m g(z)$, it follows that $f(z) \neq 0$ for all z in $D'(c, \delta)$ □

Before stating our next theorem, let us look again at the function $z \mapsto z^n$. It maps 0 to 0, and maps each neighbourhood of 0 onto another neighbourhood of 0. In fact the function maps the neighbourhood $N(0, \epsilon)$ onto $N(0, \epsilon^n)$, and for $0 < r < \epsilon$, maps each of the points $re^{i(\alpha + 2k\pi)/n}$ $(k = 0, 1, \ldots, n - 1)$ to the single point $r^n e^{i\alpha}$. We say that the neighbourhood $N(0, \epsilon)$ maps to $N(0, \epsilon^n)$ in an "n-to-one" fashion.

The following theorem in effect states that every non-constant holomorphic function behaves in essentially the same way:

Theorem 10.6

Let f be non-constant and holomorphic in a neighbourhood of c, and let $f(c) = d$. Let $g(z) = f(z) - d$, and let n be the order of the zero of g at c. If $\epsilon > 0$ is sufficiently small, then there exists $\delta > 0$ such that, for each w in $D'(d, \delta)$, there exist n distinct points z_i $(i = 1, \ldots, n)$ in $D'(c, \epsilon)$ such that $f(z_i) = w$.

Proof

By Theorem 10.5, if $\epsilon > 0$ is sufficiently small then $f(z) \neq d$ for all z in $D'(c, 2\epsilon)$. If $f'(c) \neq 0$ then the continuity of f' ensures that $f'(z) \neq 0$ in some neighbourhood of c. On the other hand, if $f'(c) = 0$ it follows by Theorem 10.5 that $f'(z) \neq 0$ in a suitably small punctured disc with centre c. So we refine our choice of ϵ so as to have both $f(z) \neq d$ and $f'(z) \neq 0$ for all z in $D'(c, 2\epsilon)$.

Since the set $\kappa(c, \epsilon) = \{z : |z - c| = \epsilon\}$ is closed and bounded, it follows by Theorem 5.3 that

$$\inf \{|g(z)| : |z - c| = \epsilon\} = \delta > 0.$$

Let $w \in D'(d, \delta)$. Then

$$f(z) - w = g(z) + (d - w).$$

Since $|g(z)| > \delta > |d - w|$ for all points on the circle $\kappa(c, \epsilon)$, and since g has a zero of order n inside that circle, it follows by Rouché's Theorem (Theorem 10.3) that $f(z) - w \, (= g(z) + (d - w))$ has n zeros z_j $(j = 1, 2, \ldots, n)$ in $D'(c, \epsilon)$. Since $g'(z) = f'(z) \neq 0$ for all z in $D'(c, \epsilon)$, these zeros are all simple, and so distinct.   □

One consequence of this result is the **Open Mapping Theorem**:

Theorem 10.7 (The Open Mapping Theorem)

Let f be holomorphic and non-constant in an open set U. Then $f(U)$ is open.

Proof

Let $d \in f(U)$, and let $c \in U$ be such that $f(c) = d$. Choose $\epsilon > 0$ and δ as in the proof of Theorem 10.6 and such that $N(c, \epsilon) \subset U$. Let $w \in N(d, \delta)$. Then there is at least one zero z_0 of $f(z) - w$ in $N(c, \epsilon)$. That is, there exists at least one z_0 in $N(c, \epsilon)$ such that $f(z_0) = w$. Thus

$$N(d, \delta) \subseteq \{f(z) \,:\, z \in N(c, \epsilon)\} \subseteq f(U).$$

Hence $f(U)$ is open. $\square$

The **Maximum Modulus Theorem** now follows easily:

Theorem 10.8 (The Maximum Modulus Theorem)

Let f be holomorphic in a domain containing $\mathrm{I}(\gamma) \cup \gamma^*$, and let $M = \sup\{|f(z)| \,:\, z \in \mathrm{I}(\gamma) \cup \gamma^*\}$. Then $|f(z)| < M$ for all z in $\mathrm{I}(\gamma)$, unless f is constant, in which case $|f(z)| = M$ throughout $\mathrm{I}(\gamma) \cup \gamma^*$.

Proof

Let c be an element of the open set $\mathrm{I}(\gamma)$, and let $N(c, \epsilon) \subset \mathrm{I}(\gamma)$. Then, by Theorem 10.7, $d = f(c)$ lies in an open set U, the image of $N(c, \epsilon)$, and U is wholly contained in the image of the function f. Hence there is a neighbourhood $N(d, \delta)$ of d contained in U and so certainly contained in the image of f, and within this neighbourhood there are certainly points w such that $|w| > |d|$. Hence, unless f is constant, the maximum value of $f(z)$ for z in $\mathrm{I}(\gamma) \cup \gamma^*$ is attained on the boundary. $\square$

Another consequence is the **Inverse Function Theorem**:

Theorem 10.9 (The Inverse Function Theorem)

Suppose that f is holomorphic in an open set containing c and that $f'(c) \neq 0$. Then there exists $\eta > 0$ such that f is one-to-one on $N = N(c, \eta)$. Let g be the inverse function of $f|_N$ (the restriction of f to N). Let $z \in N$ and write $f(z) = w$. Then $g'(w) = 1/f'(z)$.

Proof

Write $f(c)$ as d, and let ϵ, δ be as in the proof of Theorem 10.6. Since f is continuous at c, there exists η such that $0 < \eta \leq \epsilon$ and $|f(z) - d| < \delta$ whenever $|z - c| < \eta$. Let $N = N(c, \eta)$. Since $f'(c) \neq 0$, the zero of $f(z) - d$ at c is of order one, and so, by Theorem 10.6, $f|_N$ is one-to-one, and has an inverse function g.

Look again at Theorem 10.6, in the case where $n = 1$. It establishes the existence of a *single* point z_1 in $D'(c, \epsilon)$ such that $f(z_1) = w$, and we can of course write $z_1 = g(w)$. Thus $g(w) \in D'(c, \epsilon)$ whenever $w \in D'(d, \delta)$. It follows that g is continuous at d.

Finally, if $z, \zeta \in N$ and if $f(z) = w$, $f(\zeta) = \omega$, then, by Exercise 4.4,

$$f(\zeta) - f(z) = A(z)(\zeta - z),$$

where $A(z)$ is continuous at z, and tends to $f'(z)$ as $\zeta \to z$. That is,

$$\omega - w = A\big(g(w)\big)\big(g(\omega) - g(w)\big),$$

or, equivalently,

$$g(\omega) - g(w) = \frac{1}{A\big(g(w)\big)}(\omega - w).$$

Since $1/(A \circ g)$ is continuous at w, we deduce that g is differentiable at w and that $g'(w) = 1/f'\big(g(w)\big) = 1/f'(z)$. $\qquad \square$

It is important to note that the one-to-one property in the statement of the inverse function theorem is a *local* property, holding within a neighbourhood: for example, the exponential function has non-zero derivative at every point, but the function is *not* one-to-one. In fact, one-to-one entire functions are very rare:

Theorem 10.10

Let f be a non-constant entire function, one-to-one throughout $\mathbb{C}$. Then f is linear, that is, there exist $a, b \in \mathbb{C}$ such that $f(z) = az + b$ $(z \in \mathbb{C})$.

Proof

Suppose first that f is not a polynomial. By Theorem 10.7 the image of the open set $N(0, 1)$ is an open set, and so contains a neighbourhood $N(f(0), \epsilon)$ of

$f(0)$. On the other hand, from Theorem 8.11 we know that there exists z such that $|z| > 1$ and such that $|f(z) - f(0)| < \epsilon$, and so we have a contradiction to our assumption that f is one-to-one. Hence f must be a polynomial, of degree n (say), and by the Fundamental Theorem of Algebra we must have

$$f(z) = c(z - \alpha_1)(z - \alpha_2) \ldots (z - \alpha_n).$$

The one-to-one property forces all the roots to coincide, and so $f(z) = c(z-\alpha)^n$, for some α in $\mathbb{C}$ and some $n \geq 1$. Now, if ω_1, ω_2 are two distinct nth roots of 1, we have $f(\alpha + \omega_1) = f(\alpha + \omega_2) = c$, and this gives a contradiction unless $n = 1$. □

EXERCISES

10.6. Let f be holomorphic in a domain containing $\bar{N}(0, R)$, and let M be a positive real number. Show that, if $|f(z)| > M$ for all z on the circle $\kappa(0, R)$ and $|f(0)| < M$, then f has at least one zero in $N(0, R)$.

Use this result to outline a proof of the Fundamental Theorem of Algebra.

10.7. Let f be holomorphic in the closed disc $\bar{N}(0, R)$. Show that Re f cannot have a maximum value in $N(0, R)$. Can it have a minimum value? [Hint: consider e^f.]

10.3 Winding Numbers

Many of the theorems we have stated and proved in this book concerning piecewise smooth functions $\gamma(t)$ giving rise to simple, closed curves can, with some modification, be extended to curves that are not simple. The key is the notion of the **winding number** of a curve. For clarity, let us refer to γ^* as a **W-contour** if γ is closed and piecewise smooth.

If in (10.2) we take $f(z) = z - c$, then $f'(z) = 1$ and so

$$\frac{1}{i} \int_\gamma \frac{dz}{z - c} = \Delta_\gamma \big(\arg(z - c) \big).$$

If the contour γ is simple, then this is equal to 2π if $c \in \mathrm{I}(\gamma)$ and 0 if $c \in \mathrm{E}(\gamma)$, but if γ^* is a W-contour we may obtain $2n\pi$, where n is an integer other than

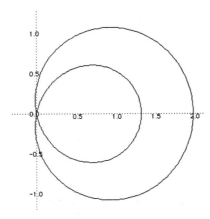

Figure 10.1. A curve that is closed but not simple

0 or 1. This integer is called the **winding number** of the W-contour, and is denoted by $w(\gamma, c)$. That is,

$$w(\gamma, c) = \frac{1}{2\pi i} \int_\gamma \frac{dz}{z - c}.$$ (10.4)

For the contour in Figure 10.1 we have $w(\gamma, 1) = 2$, $w(\gamma, 3/2) = 1$ and $w(\gamma, 3) = 0$.

It is possible to extend several key theorems to the case of W-contours. See [3] or [10]. In particular, the residue theorem becomes

Theorem 10.11

Let γ be closed and piecewise smooth, and let the function f be a meromorphic within a disc containing γ^*, with poles at $c_1, c_2, \ldots, c_n$. Then

$$\int_\gamma f(z)\,dz = 2\pi i \sum_{i=1}^n w(\gamma, c_i) \mathrm{res}(f, c_i).$$

For example, if γ^* is as in Figure 10.1, and if f is meromorphic in the open disc $N(0, 4)$ with poles at 1, $3/2$ and 3, then

$$\int_\gamma f(z)\,dz = 2\pi i [2\mathrm{res}(f, 1) + \mathrm{res}(f, 3/2) + 0\mathrm{res}(f, 3)].$$

In the same way, Theorem 10.1 becomes

Theorem 10.12

Let γ be a W-contour, let f be meromorphic in a disc D that contains γ^*, and suppose that $Q = \{q \in D : \mathrm{ord}(f, q) \neq 0\}$ is finite. Then

$$\frac{1}{2\pi i} \int_\gamma \frac{f'(z)}{f(z)}\, dz = \sum_{q \in Q} w(\gamma, q)\mathrm{ord}(f, q)\,.$$

Remark 10.13

It is of course possible that for some of the members q of the set Q we have $w(\gamma, q) = 0$. For example, referring again to Figure 10.1, we might have $f(z) = (z - 1)^2(2z - 3)(z - 3)$, with zeros at 1 (double), $3/2$ and 3, and our theorem would give

$$\int_\gamma \frac{f'(z)}{f(z)}\, dz = 2\pi i[(2 \times 2) + (1 \times 1) + (0 \times 1)] = 10\pi\,.$$

11
Conformal Mappings

11.1 Preservation of Angles

This chapter explores the consequence of a remarkable geometric property of holomorphic functions. Look again at Figures 3.1 and 3.2 on page 42. For arbitrary k and l the lines $u = k$ and $v = l$ in the w-plane are of course mutually perpendicular, and visually at least it seems that the corresponding hyperbolic curves $x^2 - y^2 = k$ and $2xy = l$ in the z-plane are also perpendicular. Again, the lines $x = k$ and $y = l$ are mutually perpendicular, and it appears also that the corresponding parabolic curves in the w-plane are also perpendicular. These observations are in fact mathematically correct (see Exercise 11.1), and are instances of a general theorem to be proved shortly. First, however, we need to develop a little more of the theory of the parametric representation of curves that was introduced in Section 5.2.

In the space $\mathbb{R}^2$ of two dimensions, the parametric representation of a straight line L through $\mathbf{a} = (a_1, a_2)$ in the direction of the non-zero vector $\mathbf{v} = (v_1, v_2)$ is

$$L = \{\mathbf{a} + t\mathbf{v} : t \in \mathbb{R}\} = \{(a_1 + tv_1, a_2 + tv_2) : t \in \mathbb{R}\}.$$

Suppose now that we have a curve

$$\mathcal{C} = \{(r_1(t), r_2(t)) : t \in [a, b]\},$$

where r_1 and r_2 are differentiable. For each u in $[a, b]$, the tangent T_c to $\mathcal{C}$ at the point $\big(r_1(u), r_2(u)\big)$ is in the direction of the vector $\big(r_1'(u), r_2'(u)\big)$, and so

$$T_u = \big\{\big(r_1(u) + tr_1'(u), r_2(u) + tr_2'(u)\big) \, : \, t \in \mathbb{R}\big\}.$$

As in Section 5.2, we can easily translate the vector $\big(r_1(t), r_2(t)\big)$ into a complex number $\gamma(t)$. Thus, if γ is differentiable, the tangent to the curve

$$\mathcal{C} = \{\gamma(t) \, : \, t \in [a,b]\}$$

at the point $\gamma(u)$ is

$$T_u = \{\gamma(u) + t\gamma'(u) \, : \, t \in \mathbb{R}\},$$

provided $\gamma'(u) \neq 0$. If $\gamma'(u) = 0$ then there is no well-defined tangent at the point $\gamma(u)$. (For example, consider the cycloid in Figure 5.1, where $\gamma'(t) = 1 - e^{-it} = 0$ when $t = 2n\pi$ $(n \in \mathbb{Z})$. At each of these points the graph has a cusp.)

Now consider two smooth curves

$$\mathcal{C}_1 = \{\gamma_1(t) \, : \, t \in [0,1]\}, \quad \mathcal{C}_2 = \{\gamma_2(t) \, : \, t \in [0,1]\},$$

intersecting in the point $\gamma_1(0) = \gamma_2(0)$. Suppose that $\gamma_1'(0)$ and $\gamma_2'(0)$ are both non-zero, so that there are well defined tangents T_1 and T_2 at the point of intersection:

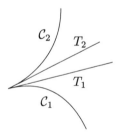

We then *define* the angle between the curves $\mathcal{C}_1$ and $\mathcal{C}_2$ to be the angle between the tangents, namely $\arg\big(\gamma_2'(0) - \gamma_1'(0)\big)$. This is a reasonable definition: for $i = 1, 2$, $\arg\big(\gamma_i'(0)\big)$ is the angle made by the non-zero vector $\gamma_i'(0)$ with the positive x-axis, and the angle between the two vectors $\gamma_1'(0)$ and $\gamma_2'(0)$ is the difference between the two arguments.

We now have a theorem which says roughly that angles are preserved by holomorphic functions. More precisely, if we take account of the potential ambiguity in the argument, we have

Theorem 11.1

Let f be holomorphic in an open subset U of $\mathbb{C}$. Suppose that two curves

$$\mathcal{C}_1 = \{\gamma_1(t) : t \in [0,1]\}, \quad \mathcal{C}_2 = \{\gamma_2(t) : t \in [0,1]\},$$

lying inside U meet at a point $c = \gamma_1(0) = \gamma_2(0)$. Suppose that $f'(c)$, $\gamma_1'(0)$ and $\gamma_2'(0)$ are all non-zero. Let

$$\mathcal{D}_1 = \{f(\gamma_1(t)) : t \in [0,1]\}, \quad \mathcal{D}_2 = \{f(\gamma_2(t)) : t \in [0,1]\}.$$

If the angle between $\mathcal{C}_1$ and $\mathcal{C}_2$ is ϕ and the angle between $\mathcal{D}_1$ and $\mathcal{D}_2$ is ψ, then $\psi \equiv \phi \pmod{2\pi}$.

Proof

The curves $\mathcal{D}_1$ and $\mathcal{D}_2$ meet at $f(c)$ at an angle

$$\psi = \arg(f \circ \gamma_2)'(0) - \arg(f \circ \gamma_2)'(0).$$

By the chain rule,

$$\frac{(f \circ \gamma_1)'(0)}{(f \circ \gamma_2)'(0)} = \frac{f'(\gamma_1(0))\gamma_1'(0)}{f'(\gamma_2(0))\gamma_2'(0)} = \frac{f'(c)\gamma_1'(0)}{f'(c)\gamma_2'(0)} = \frac{\gamma_1'(0)}{\gamma_2'(0)}. \tag{11.1}$$

Hence

$$\begin{aligned}
\psi &= \arg(f \circ \gamma_2)'(0) - \arg(f \circ \gamma_2)'(0) \\
&\equiv \arg \gamma_2'(0) - \arg \gamma_2'(0) \pmod{2\pi} \quad \text{(by (11.1) and Exercise 2.6)} \\
&= \phi.
\end{aligned}$$

$\square$

Remark 11.2

The proof of the theorem makes it clear that the **sense** as well as the magnitude of the angle between $\mathcal{C}_1$ and $\mathcal{C}_2$ is preserved by f. An obvious example of a (non-holomorphic) f preserving magnitude but not sense is $f : z \mapsto \bar{z}$, which we can think of geometrically as reflection in the x-axis.

Remark 11.3

The trivial observation that

$$\frac{|f(z) - f(c)|}{|z - c|} \to |f'(c)| \quad \text{as } z \to c$$

has a geometric interpretation, that the **local magnification** of the mapping f at the point c is $|f'(c)|$.

Remark 11.4

The condition $f'(c) \neq 0$ in the statement of Theorem 11.1 is essential. For example, consider the holomorphic function $f : z \mapsto z^2$, noting that $f'(0) = 0$. The positive x-axis maps to itself, and the line $\theta = \pi/4$ maps to the positive y-axis. The angle between the lines doubles.

We shall say that a complex function f is **conformal** in an open set U if it is holomorphic in U and if $f'(c) \neq 0$ for all c in U. Thus, for example, the function $z \mapsto z^2$ is conformal in the open set $\mathbb{C} \setminus \{0\}$. Theorem 11.1 tells us that conformal mappings preserve angles.

EXERCISES

11.1 With reference to Figures 3.1 and 3.2, suppose that $k, l, x, y \neq 0$.

    a) Show that the hyperbolas $x^2 - y^2 = k$ and $2xy = l$ (with $k, l > 0$) meet at right angles.

    b) Show that the parabolas $v^2 = 4k^2(k^2 - u)$ and $v^2 = 4l^2(l^2 + u)$ meet at right angles.

11.2 Let f be conformal in the open set U. Show that the function $g : z \mapsto f(\bar{z})$ preserves the magnitude of angles but not the sense.

11.2 Harmonic Functions

Let U be an open subset of $\mathbb{R}^2$. A function $f : U \to \mathbb{R}$ is said to be **harmonic** if

(i) f has continuous second order partial derivatives in U;

(ii) f satisfies **Laplace's**[1] **equation**

$$\frac{\partial^2 f}{\partial x^2} + \frac{\partial^2 f}{\partial y^2} = 0. \tag{11.2}$$

[1] Pierre Simon Laplace, 1749–1827.

Harmonic functions are of immense importance in applied mathematics, and one major reason why complex analysis is important to applied mathematicians is that harmonic and holomorphic functions are closely related. First, we have

Theorem 11.5

Let f be holomorphic in an open set U, with real and imaginary parts u and v. Then both u and v are harmonic in U.

Proof

We are supposing that $u, v : \mathbb{R}^2 \to \mathbb{R}$ are such that

$$f(x + iy) = u(x, y) + iv(x, y).$$

Since f is infinitely differentiable by Theorem 7.5, we know that u and v have partial derivatives of all orders, and by the Cauchy–Riemann equations we have

$$\frac{\partial^2 u}{\partial x^2} = \frac{\partial}{\partial x}\left(\frac{\partial u}{\partial x}\right) = \frac{\partial}{\partial x}\left(\frac{\partial v}{\partial y}\right) = \frac{\partial}{\partial y}\left(\frac{\partial v}{\partial x}\right) = \frac{\partial}{\partial y}\left(-\frac{\partial u}{\partial y}\right) = -\frac{\partial^2 u}{\partial y^2},$$

and similarly

$$\frac{\partial^2 v}{\partial x^2} = -\frac{\partial^2 v}{\partial y^2}.$$

Thus both u and v are harmonic functions. □

There is a converse:

Theorem 11.6

Let D be an open disc, and suppose that $u : D \to \mathbb{R}$ is harmonic. Then there exists a complex function f, holomorphic in D, such that $u = \operatorname{Re} f$.

Proof

If such an f exists, with $\operatorname{Re} f = u$ and $\operatorname{Im} f = v$ (say), then $f'(z) = u_x + iv_x = u_x - iu_y$. So we define $g(z) = u_x(x, y) - iu_y(x, y)$ (where, as usual, $z = x + iy$). Then u_x and $-u_y$ have continuous first order partial derivatives and, in D, the Cauchy–Riemann equations are satisfied by the real and imaginary parts of g:

$$(u_x)_x = (-u_y)_y, \quad (u_x)_y = -(-u_y)_x.$$

Hence, by Theorem 4.3, g is holomorphic in U. By Theorem 5.25 there exists a holomorphic function G such that $G' = g$. If we write $\operatorname{Re} G = H$, $\operatorname{Im} G = K$, then

$$G' = H_x + iK_x = H_x - iH_y = u_x - iu_y \,,$$

and so $(H - u)_x = (H - u)_y = 0$ throughout D. Hence $H(x, y) - u(x, y) = k$, a real constant. Let $f(z) = G(z) - k$; then $\operatorname{Re} f = H - k = u$. $\qquad\square$

Remark 11.7

The function $\operatorname{Im} f$, which is also harmonic, is called a **harmonic conjugate** for u.

In practice a certain amount of guessing can produce a harmonic conjugate and an associated holomorphic function:

Example 11.8

Let $u(x, y) = x^3 - 3xy^2 - 2y$. Verify that u is harmonic, and determine a function v such that $f = u + iv$ is holomorphic.

Solution

We easily see that

$$\frac{\partial^2 u}{\partial x^2} + \frac{\partial^2 u}{\partial y^2} = 6x + (-6x) = 0 \,.$$

The required function v must satisfy the Cauchy–Riemann equations, and so

$$\frac{\partial v}{\partial y} = \frac{\partial u}{\partial x} = 3x^2 - 3y^2 \,.$$

By integration we deduce that $v(x, y) = 3x^2y - y^3 + g(x)$ for some function g. Then from

$$\frac{\partial v}{\partial x} = 6xy + g'(x) \quad \text{and} \quad -\frac{\partial u}{\partial y} = 6xy + 2$$

we deduce from the Cauchy–Riemann equations that $g'(x) = 2$. Hence, choosing $g(x) = 2x$, we obtain

$$v(x, y) = 3x^2y - y^3 + 2x \,,$$

and we easily verify that v is again a harmonic function. Observe that

$$f(z) = (x^3 - 3xy^2 - 2y) + i(3x^2y - y^3 + 2x) = z^3 + 2iz \,.$$

$\qquad\square$

Many boundary-value problems in applied mathematics come under the general heading of the **Dirichlet[2] problem**:

---

[2] Johann Peter Gustav Lejeune Dirichlet, 1805–1859.

- Let U be an open set bounded by a simple closed piecewise smooth curve, and let F be a continuous real-valued function with domain ∂U, the boundary of U. Can we find a function f that is continuous on $\bar{U}$, harmonic in U, and such that $f = F$ on ∂U?

The following solution (which I shall not prove) exists for the case where U is the open disc $N(0,1)$. Here the boundary value function F has domain $\{e^{i\theta} : 0 \le \theta < 2\pi\}$. For $re^{i\theta}$ in U, let

$$g(re^{i\theta}) = \frac{1}{2\pi} \int_0^{2\pi} \frac{1 - r^2}{1 - 2r\cos(\theta - t) + r^2} F(e^{it})\, dt. \qquad (11.3)$$

Let

$$f(re^{i\theta}) = \begin{cases} g(re^{i\theta}) & \text{if } 0 \le r < 1 \\ F(re^{i\theta}) & \text{if } r = 1; \end{cases}$$

then f is harmonic in $N(0,1)$, continuous in $\bar{N}(0,1)$, and $f = F$ on the circle $\kappa(0,1)$. (For a proof, see [4].)

It is clear that rescaling and translating will give a solution of the Dirichlet problem for a general open disc $N(a, R)$. We now look at a strategy for solving the problem in the general case. The first element of the strategy is a theorem due to Riemann:

Theorem 11.9 (The Riemann Mapping Theorem)

Let γ^* be a contour. Then there exists a one-to one conformal mapping f from $\mathrm{I}(\gamma)$ onto $N(0,1)$, with $f^{-1} : N(0,1) \to \mathrm{I}(\gamma)$ also conformal.

A proof of this can be found in [4]. Unfortunately there is no practical general method for finding the function f, which is why the next section will deal with a number of ways of transforming open sets using conformal mappings.

The other part of the strategy is a theorem to the effect that the composition of a harmonic function and a conformal mapping is harmonic. Precisely,

Theorem 11.10

Let $D = \mathrm{I}(\gamma)$, where γ defines a simple, closed, piecewise smooth curve in the z-plane, and let f, where $f(x+iy) = u(x,y)+iv(x,y)$, be a conformal mapping transforming D into D^* (in the w-plane). If $\theta^*(u,v)$ is harmonic in D^*, then θ, given by

$$\theta(x,y) = \theta^*\big(u(x,y), v(x,y)\big),$$

is harmonic in D.

Proof

By Theorem 11.6, there is a harmonic conjugate ϕ^* for θ^* such that $F^*(w) = \theta^*(u,v) + i\phi^*(u,v)$ is holomorphic in D^*. Hence $F^* \circ f$ is holomorphic in D, and so θ, its real part, is harmonic in D.   □

To "solve" the Dirichlet problem we first use a suitable holomorphic function f to transform U into $N(0,1)$, find the appropriate harmonic function in $N(0,1)$, then use f^{-1} to transform back into U. This is a strategy rather than a solution, for there are practical difficulties in the way. As already mentioned, there is no general method for finding f, and the feasibility of evaluating the integral in (11.3) depends on the nature of the boundary value function F.

Using the close connection with holomorphic functions, we finish this section by establishing a maximum principle for harmonic functions.

Theorem 11.11

Let γ be a contour, and let u be harmonic and non-constant in $I(\gamma) \cup \gamma^*$. Then u is bounded in $I(\gamma) \cup \gamma^*$. Let

$$M = \sup \{u(x,y) : (x,y) \in I(\gamma) \cup \gamma^*\}.$$

Then $u(x,y) < M$ for all (x,y) in $I(\gamma)$.

Proof

Choose v so that $f = u + iv$ is holomorphic. Then $g : z \mapsto e^{f(z)}$ is again holomorphic. Observe that

$$|g(z)| = e^{u(x,y)}. \tag{11.4}$$

By Theorem 5.3, $g(z)$ is bounded, and it follows from (11.4) that $u(x,y)$ is bounded above. Next, observe that

$$e^M = \sup \{|g(z)| : z \in I(\gamma) \cup \gamma^*\}.$$

By the Maximum Modulus Theorem (Theorem 10.8), $|g(z)| < e^M$ for all z in $I(\gamma)$. Hence $u(x,y) < M$ for all (x,y) in $I(\gamma)$.   □

EXERCISES

11.3. Verify that the following functions u are harmonic, and determine a function v such that $u + iv$ is a holomorphic function.

a) $u(x,y) = x(1 + 2y)$;

b) $u(x,y) = e^x \cos y$;

c) $u(x,y) = x - \dfrac{y}{x^2 + y^2}$.

11.4 With reference to Theorem 11.11, let

$$m = \inf\left\{u(x,y) \,:\, (x,y) \in \mathrm{I}(\gamma) \cup \gamma^*\right\}.$$

Show that $u(x,y) > m$ for all (x,y) in $\mathrm{I}(\gamma)$.

11.3 Möbius Transformations

In transforming regions by means of conformal mappings, straight lines and circles play an significant role, and there is an important class of conformal mappings that transform circles and lines into circles and lines.

In this section it is convenient to deal with the extended complex plane $\mathbb{C} \cup \{\infty\}$ mentioned in Section 3.3. (Recall that there is a *single* point at infinity.) We shall denote $\mathbb{C} \cup \{\infty\}$ by $\mathbb{C}^*$, and will use the following conventions, in which c is a (finite) complex number:

$$c \pm \infty = \pm\infty + c = \infty \quad c \times \infty = \infty \times c = \infty$$

$$c/\infty = 0\,, \quad c/0 = \infty\,.$$

$$\overline{\infty} = -\infty = \infty\,, \quad \infty + \infty = \infty \times \infty = \infty\,.$$

We shall also extend the meaning of "circle" to include a straight line, which we think of as a circle with infinite radius.

A **Möbius**[3] transformation, also called a **bilinear** transformation, is a map

$$z \mapsto \frac{az + b}{cz + d} \qquad (z \in \mathbb{C}^*\,,\ a,b,c,d \in \mathbb{C}\ \ ad - bc \neq 0)\,. \tag{11.5}$$

The condition $ad - bc \neq 0$ is necessary for the transformation to be of interest: if $a = b = c = d = 0$ the formula is meaningless, and otherwise the condition $ad - bc = 0$ gives $a/c = b/d = k$ (say) and the transformation reduces to $z \mapsto k$.

It is clear that a Möbius transformation is holomorphic except for a simple pole at $z = -d/c$. Its derivative is the function

$$z \mapsto \frac{ad - bc}{(cz + d)^2}\,,$$

and so the mapping is conformal throughout $\mathbb{C} \setminus \{-d/c\}$.

---

[3] August Ferdinand Möbius 1790–1868.

Theorem 11.12

The inverse of a Möbius transformation is a Möbius transformation. The composition of two Möbius transformations is a Möbius transformation.

Proof

As is easily verified, the Möbius transformation

$$w \mapsto \frac{dw - b}{-cw + a} \tag{11.6}$$

is the inverse of $z \mapsto (az + b)/(cz + d)$.

Given Möbius transformations

$$f_1 : z \mapsto \frac{a_1 z + b_1}{c_1 z + d_1} \quad \text{and} \quad f_2 : z \mapsto \frac{a_2 z + b_2}{c_2 z + d_2},$$

an easy calculation gives

$$(f_1 \circ f_2)(z) = \frac{Az + B}{Cz + D},$$

where

$$A = a_1 a_2 + b_1 c_2, \quad B = a_1 b_2 + b_1 d_2, \quad C = c_1 a_2 + d_1 c_2, \quad D = c_1 b_2 + d_1 d_2.$$

Thus $f_1 \circ f_2$ is a Möbius transformation, since a routine calculation gives

$$AD - BC = (a_1 d_1 - b_1 c_2)(a_2 d_2 - b_2 c_2) \neq 0.$$

$\square$

Remark 11.13

The composition of Möbius transformations in effect corresponds to matrix multiplication. If we define the matrices of f_1 and f_2 as

$$\begin{pmatrix} a_1 & b_1 \\ c_1 & d_1 \end{pmatrix} \quad \text{and} \quad \begin{pmatrix} a_2 & b_2 \\ c_2 & d_2 \end{pmatrix},$$

then the matrix of $f_1 \circ f_2$ is

$$\begin{pmatrix} a_1 & b_1 \\ c_1 & d_1 \end{pmatrix} \begin{pmatrix} a_2 & b_2 \\ c_2 & d_2 \end{pmatrix}.$$

Recalling that

$$\begin{pmatrix} a & b \\ c & d \end{pmatrix}^{-1} = \frac{1}{ad - bc} \begin{pmatrix} d & -b \\ -c & a \end{pmatrix},$$

we see that this is essentially the matrix of the inverse of $z \mapsto (az+b)/(cz+d)$ as indicated in (11.6), since multiplication of all the coefficients by a non-zero complex constant k does not change a Möbius transformation.

Among special Möbius transformations are:

(M1) $z \mapsto az$ $(b = c = 0, d = 1)$;

(M2) $z \mapsto z + b$ $(a = d = 1, c = 0$ – translation by b);

(M3) $z \mapsto 1/z$ $(a = d = 0, b = c = 1$ – inversion).

In (M1), if $a = Re^{i\theta}$, the geometrical interpretation is an expansion by the factor R followed by a rotation anticlockwise by θ.

Theorem 11.14

Every Möbius transformation

$$F : z \mapsto \frac{az+b}{cz+d}$$

is a composition of transformations of type (M1), (M2) and (M3).

Proof

If $c = 0$ then $d \neq 0$ and it is clear that $F = g_2 \circ g_1$, where

$$g_1 : z \mapsto \frac{a}{d} z, \quad g_2 : z \mapsto z + \frac{b}{d}.$$

If $c \neq 0$, then $F = g_5 \circ g_4 \circ g_3 \circ g_2 \circ g_1$, with

$$g_1 : z \mapsto cz, \quad g_2 : z \mapsto z + d, \quad g_3 : z \mapsto \frac{1}{z},$$

$$g_4 : z \mapsto \frac{1}{c}(bc - ad)z, \quad g_5 : z \mapsto z + \frac{a}{c},$$

for

$$g_1(z) = cz, \quad (g_2 \circ g_1)(z) = cz + d, \quad (g_3 \circ g_2 \circ g_1)(z) = \frac{1}{cz + d},$$

$$(g_4 \circ g_2 \circ g_1)(z) = \frac{bc - ad}{c(cz + d)}, \quad (g_5 \circ g_4 \circ g_2 \circ g_1)(z) = \frac{a}{c} + \frac{bc - ad}{c(cz + d)} = \frac{az + b}{cz + d}.$$

$\square$

It is clear that transformations of type (M1) and (M2) preserve shapes, and in particular that they transform circles to circles. Inversion transformations in general will change shapes, but circles survive. From Theorem 2.9, a circle Σ in the z-plane can be written as the set of z such that

$$\left| \frac{z-c}{z-d} \right| = k \,,$$

where $c, d \in \mathbb{C}$ and $k > 0$. If $w = 1/z$, the image Σ' in the w-plane is the set of w such that

$$k = \left| \frac{(1/w)-c}{(1/w)-d} \right| = \frac{|c|}{|d|} \left| \frac{w-(1/c)}{w-(1/d)} \right| \,, \tag{11.7}$$

and so this too is a circle. Recall from Remark 2.11 that the points c and d are inverse points with respect to the circle Σ. From (11.7) it follows that their images $1/c$ and $1/d$ are inverse points with respect to Σ'.

It thus follows from Theorem 11.14 that (with our extended definition of "circle") we have:

Theorem 11.15

A Möbius transformation transforms circles into circles, and inverse points into inverse points.

Here we must recall that two points are **inverse** with respect to a line if each is the reflection of the other in the line.

If we require to find a Möbius transformation that carries out a particular transformation, it can be useful to know a way of writing down the transformation that sends three chosen distinct points to distinct chosen destinations:

Theorem 11.16

Let (z_1, z_2, z_3), (w_1, w_2, w_3) be triples of distinct points. There is a unique Möbius transformation f mapping z_i to w_i for $i = 1, 2, 3$.

Proof

The Möbius transformation

$$g : z \mapsto \left(\frac{z-z_1}{z-z_3} \right) \left(\frac{z_2-z_3}{z_2-z_1} \right)$$

maps z_1, z_2, z_3 to 0, 1, ∞, respectively. Similarly, the Möbius transformation

$$h : w \mapsto \left(\frac{w-w_1}{w-w_3} \right) \left(\frac{w_2-w_3}{w_2-w_1} \right)$$

maps w_1, w_2, w_3 to 0, 1, ∞. It follows that $h^{-1} \circ g$ maps z_1, z_2, z_3 into w_1, w_2, w_3.

To prove uniqueness, suppose first that $j : z \mapsto (az + b)/(cz + d)$ has 0, 1, ∞ as fixed points. Then $j(0) = 0$ implies that $b = 0$, $j(\infty) = \infty$ implies that $c = 0$, and $j(1) = 1$ implies that $a = d$. It follows that j is the identity function $z \mapsto z$. If p is a Möbius transformation mapping each z_i to w_i, then $h \circ p \circ g^{-1}$ fixes 0, 1, ∞, and so is equal to the identity function j. It follows that

$$p = (h^{-1} \circ h) \circ p \circ (g_{-1} \circ g) = h^{-1} \circ (h \circ p \circ g^{-1}) \circ g = h^{-1} \circ j \circ g = h^{-1} \circ g.$$

$\square$

From Theorems 11.15 and 11.16 we deduce:

Theorem 11.17

Let S_1, S_2 be circles in the plane. There exists a Möbius transformation mapping S_1 onto S_2.

Proof

A circle is determined by the position of three distinct points on it. (If the three points are collinear, or if one of the points is ∞ we have a straight line.) Choose three distinct points on S_1 and three on S_2. A suitable Möbius transformation is then the unique Möbius transformation F mapping the three chosen points on S_1 to the three chosen points on S_2. $\square$

Example 11.18

Find a Möbius transformation mapping the real axis $L = \{z : \operatorname{Im} z = 0\}$ onto the circle $S = \{z : |z| = 1\}$.

Solution

Choose $-1, 0, 1$ on L and $-1, i, 1$ on S. Let $F(z) = (az + b)/(cz + d)$, and suppose that $F(-1) = -1$, $F(0) = i$, $F(1) = 1$. We obtain the equations

$$-a + b = c - d, \quad b = id, \quad a + b = c + d,$$

from which we easily deduce that

$$F(z) = \frac{z + i}{iz + 1}.$$

The solution is not unique.

We can check the answer. If z is real, then $|z+i| = \sqrt{1+z^2} = |iz+1|$, and so $|F(z)| = 1$. Also,

$$F^{-1}(w) = \frac{w-i}{-iw+1},$$

and if $|w| = 1$, then

$$F^{-1}(w) = \frac{(w-i)(1+i\bar{w})}{|1-iw|^2} = \frac{w+\bar{w}+i(|w|^2-1)}{|1-iw|^2} = \frac{2\operatorname{Re}w}{|1-iw|^2} \in \mathbb{R}.$$

$\square$

Remark 11.19

In Example 11.18 above, we can in fact obtain more information. If $\operatorname{Im}z > 0$ then z is closer to i than to $-i$, and so $|z-i| < |z+i|$. It follows that

$$|F(z)| = \left| \frac{z-i}{i(z+i)} \right| = \frac{|z-i|}{|z+i|} < 1,$$

and so the upper half-plane $\{z : \operatorname{Im}z > 0\}$ maps to the interior $\{w : |w| < 1\}$ of the circle. If we wanted the upper half-plane to map to the exterior, then

$$z \mapsto \frac{iz+1}{z+i}$$

would do the trick.

The next example shows that we can control the exterior/interior question in advance by using the inverse points property of Theorem 11.15.

Example 11.20

Find a Möbius transformation mapping the half-plane $\{z : \operatorname{Re}z \leq 1\}$ onto $\{z : |z-1| \geq 2\}$.

Solution

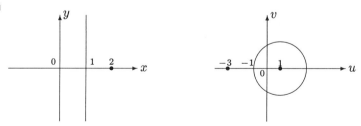

The points 0 and 2 are inverse points with respect to the line

$$L = \{z : \operatorname{Re}z = 1\}$$

while the points 0 and -3 are inverse with respect to the circle

$$S = \{w \, : \, |w - 1| = 2\}\,.$$

Since we wish to map $\{z \, : \, \mathrm{Re}\, z \leq 1\}$ to the *exterior* of the circle S, we look for a Möbius transformation F such that $F(0) = -3$ and $F(2) = 0$. We now choose 1, a point on L and -1, a point on S, and suppose that $F(1) = -1$. Writing $F(z)$ as $(az + b)/(cz + d)$, we then have

$$b = -3d\,, \quad 2a + b = 0\,, \quad a + b = -c - d\,,$$

from which we easily deduce that

$$F(z) = \frac{3z - 6}{z + 2}\,. \tag{11.8}$$

□

We can examine the transformation (11.8) more closely. Consider its inverse, given by

$$F^{-1}(z) = \frac{2z + 6}{-z + 3}\,, \tag{11.9}$$

which maps the circle $S = \{z \, : \, |z - 1| = 2\}$ onto the line $L = \{w \, : \, \mathrm{Re}\, w = 1\}$. A typical point $1 + 2e^{i\theta}$ on S maps to

$$w = \frac{2(1 + 2e^{i\theta}) + 6}{-(1 + 2e^{i\theta}) + 3} = \frac{4 + 2e^{i\theta}}{1 - e^{i\theta}} = \frac{(4 + 2e^{i\theta})(1 - e^{-i\theta})}{(1 - e^{i\theta})(1 - e^{-i\theta})}$$

$$= \frac{2 + 2e^{i\theta} - 4e^{-i\theta}}{2 - 2\cos\theta} = \frac{(2 - 2\cos\theta) + 6i\sin\theta}{2 - 2\cos\theta} = 1 + \frac{3i\sin\theta}{1 - \cos\theta}$$

$$= 1 + \frac{6i\sin(\theta/2)\cos(\theta/2)}{2\sin^2(\theta/2)} = 1 + 3i\cot\frac{\theta}{2}\,.$$

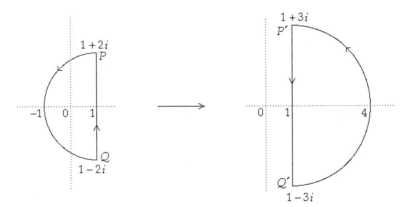

Thus, as θ goes from 0 to 2π, the image point in the w-plane traverses the line $\operatorname{Re} z = 1$ from $1 + i\infty$ to $1 - i\infty$. The image of the semicircle from $\pi/2$ to $3\pi/2$ (from the point $1 + 2i$ to the point $1 - 2i$) is the line segment from $1 + 3i$ to $1 - 3i$. To find the image of the line from Q to P, we note that $\operatorname{Re} z = 1$ if and only if $z + \bar{z} = 2$, that is (by (11.8)) if and only if

$$\frac{3w - 6}{w + 2} + \frac{3\bar{w} - 6}{\bar{w} + 2} = 2 \,.$$

After a bit of algebra, this reduces to $w\bar{w} - (w + \bar{w}) - 8 = 0$, that is, to $(u - 1)^2 + v^2 = 9$, the circle with centre 1 and radius 3. That the image is the semicircle shown can be seen either by noting that the image of 1 under the transformation (11.9) is 4, or by using the conformal property to argue as follows: a spider in the z-plane crawling from P to Q on the semicircular arc makes a left turn of $\pi/2$ when it moves on to the line-segment from Q to P; the image spider, which has been crawling on the line segment from P' to Q', must then make a left turn of $\pi/2$ on to the semicircular arc from Q' to P' shown in the diagram.

EXERCISES

11.5 Find Möbius transformations mapping:

    a) 1, i, 0 to 0, 1, ∞, respectively;

    b) 0, 1, ∞ to i, ∞, 1, respectively;

    c) 1, i, -1 to i, -1, ∞, respectively.

11.6 Determine the local magnification of the Möbius transformation $z \mapsto (az + b)/(cz + d)$ at a point ζ in $\mathbb{C}$.

11.7 a) Determine a Möbius transformation mapping the disc

$$D_1 = \{z \,:\, |z + 1| \le 2\}$$

    onto the complement of the disc $D_2 = \{z \,:\, |z + 2| < 1\}$.

    b) Determine a Möbius transformation mapping the disc D_1 (as above) onto the half-plane $\{z \,:\, \operatorname{Im} z \ge 3\}$.

11.8 Find a Möbius transformation F which maps the disc

$$D_1 = \{z \,:\, |z| \le 1\}$$

    onto the disc $D_2 = \{w \,:\, |w - 1| \le 1\}$, and such that $F(0) = \frac{1}{2}$, $F(1) = 0$. Do these properties define F uniquely?

11.9 Let
$$F(z) = \frac{1 + iz}{i + z} \, .$$

Obtain formulae for F^2 $(= F \circ F)$, F^3 and F^4. Describe the image under each of F, F^2, F^3 and F^4 of the line segment on the real axis between -1 and 1.

11.4 Other Transformations

We have already noted the geometrical aspects of the mapping $z \mapsto z^2$, which is holomorphic for all z and conformal for all $z \neq 0$. This applies more generally to the mapping $F_n : z \mapsto z^n$, where $n \geq 2$ is a positive integer. Clearly F_n maps the unit circle to itself, but the mapping is not one-to-one: the arc $\{e^{i\theta} : 0 \leq \theta < 2\pi/n\}$ maps to the whole circle $\{e^{in\theta} : 0 \leq \theta < 2\pi\}$.

Angles between curves are preserved unless they meet at the origin, where they multiply by n: if $z = e^{i\theta}$, so that Oz makes an angle θ with the positive x-axis, then $z^n = e^{in\theta}$, and Oz^n makes an angle $n\theta$ with the x-axis.

The mapping $z \to z^\alpha$, where $\alpha \geq 0$ is real, is of course a multifunction if $\alpha \notin \mathbb{N}$, but is conformal in a suitably cut plane.

Example 11.21

Let $\alpha \in (0, \pi)$. Find a transformation, conformal in
$$\{re^{i\theta} : r > 0, \, -\pi < \theta < \pi\}$$
that maps the sector $\{re^{i\theta} : r > 0, \, 0 < \theta < \alpha\}$ onto the half-plane
$$\{w : \operatorname{Im} w > 0\} \, .$$

Solution

Let $F(z) = z^{\pi/\alpha}$. Then
$$\begin{aligned} \{F(re^{i\theta}) : r > 0, \, 0 < \theta < \alpha\} &= \{r^{\pi/\alpha} e^{i\theta\pi/\alpha} : r > 0, \, 0 < \theta < \alpha\} \\ &= \{\rho e^{i\phi} : \rho > 0, \, 0 < \phi < \pi\}, \end{aligned}$$
where $\rho = r^{\pi/\alpha}$, $\phi = \theta\pi/\alpha$. □

We have already remarked that the image under $z \mapsto z^2$ of a circle $\{z : |z| = R\}$ is again a circle with centre O, but the circle in the w-plane is traversed twice. Circles with centre other than the origin have more complicated images:

Example 11.22

Find the image under $z \mapsto z^2$ of the circle $S = \{z : |z - 1| = 3\}$.

Solution

If $z = 1 + 3e^{i\theta}$ is an arbitrary point on the circle S, then

$$w = 1 + 6e^{i\theta} + 9e^{2i\theta},$$

and so

$$w + 8 = \left(6 + 9(e^{i\theta} + e^{-i\theta})\right)e^{i\theta} = 6(1 + 3\cos\theta)e^{i\theta}.$$

The path of w is a curve called a *limaçon*, and looks like this:

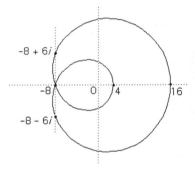

As z moves on the circle S along the upper arc from 4 to -2, w moves from the point 16 along the loop through the points $-8 + 6i$ and -8 to the point 4. Then, as z continues along the lower arc from -2 back to 4, w moves from 4, passes through -8 and $-8 - 6i$, and finishes at 16.   □

The exponential function $z \mapsto e^z = \exp z$ is conformal for all z. As we recorded in (4.25), if $z = x + iy$, then $e^z = e^x e^{iy}$, and so

$$|e^z| = e^x \qquad \arg(e^z) \equiv y \pmod{2\pi}.$$

The line $x = a$ maps by exp to the circle $|w| = e^a$, and the line $y = a$ maps to the half-line $\arg w = a$. A vertical strip bounded by $x = a$ and $x = b$ maps to the annulus

$$\{w : e^a \leq |w| \leq e^b\};$$

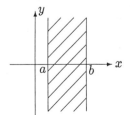 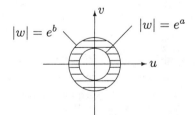

and, if $|a - b| < 2\pi$, a horizontal strip bounded by $y = a$ and $y = b$ maps to an infinite wedge between the half-lines $\arg w = a$ and $\arg w = b$:

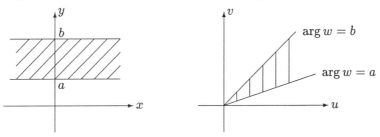

If we drop the restriction that $|a - b| < 2\pi$, this needs some qualification. Certainly it is the case that as z moves from ia to ib, w moves on the unit circle from e^{ia} to e^{ib}, but may in the process have travelled all the way round several times.

We can sometimes combine different transformations to achieve a desired geometric effect:

Example 11.23

Find a conformal mapping that transforms the sector $\{z : 0 < \arg z < \pi/4\}$ into the disc $\{w : |w - 1| < 2\}$.

Solution

First transform the sector into the upper half-plane $\{z : \text{Im } z > 0\}$ using $z \mapsto z^4$. Then find a Möbius transformation mapping the half-plane to the disc. This is not unique, but one way is to map 0 (on the half-plane) to -1 (on the circle), and to map the inverse points i and $-i$ relative to the half-plane to the inverse points 1 and ∞ relative to the circle. We obtain the Möbius transformation $z \mapsto (3z - i)/(z + i)$. The required conformal mapping is

$$z \mapsto \frac{3z^4 - i}{z^4 + i}.$$

$\square$

Example 11.24

Find a conformal mapping that transforms the vertical strip

$$S = \{z : -1 \leq \text{Re } z \leq 1\}$$

into the disc $D = \{z : |z| \leq 1\}$.

Solution

The mapping $z \mapsto iz$ transforms S into the horizontal strip

$$S_1 = \{z : -1 \leq \operatorname{Im} z \leq 1\}.$$

Then the mapping $z \mapsto e^z$ transforms S_1 to $S_2 = \{z : -1 \leq \arg z \leq 1\}$. Next, the mapping $z \mapsto z^{\pi/2}$ transforms S_2 to the half-plane $S_3 = \{z : \operatorname{Re} z \geq 0\}$. Finally, the Möbius transformation $z \mapsto (-z+1)/(z+1)$ maps S_3 to the disc D. The composition of these four mappings is

$$z \mapsto \frac{-e^{iz\pi/2} + 1}{e^{iz\pi/2} + 1}.$$

This map is holomorphic unless $e^{iz\pi/2} + 1 = 0$, that is, unless $z = 4n+2$ $(n \in \mathbb{Z})$, and this cannot happen within the strip S. Its derivative is easily seen to be

$$\frac{-i\pi e^{i\pi z/2}}{(e^{iz\pi/2} + 1)^2},$$

and this is non-zero throughout the strip S. □

Finally, we mention a class of transformations which give rise to curves known as **Joukowski's**[4] **aerofoils**. Their importance lay in the fact that they transformed circles into shapes that approximated to the profile of an aeroplane wing, and facilitated the study of the air flow round the wing. We shall look only at the simplest of this class of transformations.

The general Joukowski transformation is given by the formula

$$\frac{w - ka}{w + ka} = \left(\frac{z - a}{z + a}\right)^k.$$

We shall consider only the simplest case, when $a = 1$ and $k = 2$:

$$\frac{w - 2}{w + 2} = \left(\frac{z - 1}{z + 1}\right)^2, \tag{11.10}$$

and this simplifies to

$$w = z + \frac{1}{z}.$$

The more complicated formula (11.10) demonstrates that the Joukowski transformation is a composition $h^{-1} \circ g \circ f$, where

$$f : z \mapsto \frac{z - 1}{z + 1}, \quad g : z \mapsto z^2, \quad h : z \mapsto \frac{z - 2}{z + 1},$$

---

[4] Nikolai Egorovich Joukowski (Zhukovskiĭ), 1847–1921.

for (11.10) can be written as

$$h(w) = g\big(f(z)\big).$$

Observe that both f and g^{-1} are Möbius transformations. Accordingly, the Joukowski transformation transforms a circle first into another circle, then by squaring into a limaçon, and finally, by the Möbius transformation h^{-1}, into the aerofoil shape. The diagram below shows what happens to the circle with centre $(-1/4) + (1/2)i$ passing through the point 1.

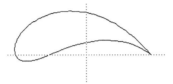

For more information on special transformations, see [5].

EXERCISES

11.10 Find the image of the first quadrant

$$Q = \{z \ : \ \mathrm{Re}\, z \geq 0\,, \ \mathrm{Im}\, z \geq 0\}$$

under the mappings

$$F_1 \ : \ z \mapsto \left(\frac{z-1}{z+1}\right)^2, \qquad F_2 \ : \ z \mapsto \frac{z^2-1}{z^2+1}\,.$$

11.11 Find a conformal mapping that transforms

$$E = \{z \ : \ \mathrm{Re}\, z \geq 0\,, \ \mathrm{Im}\, z \geq 0\,, \ |z| \geq 1\}$$

into $\{w \ : \ |w| \leq 1\}$.

11.12 Find a conformal mapping that transforms

$$Q = \{z \ : \ \mathrm{Re}\, z < \pi/2\,, \ \mathrm{Im}\, z > 0\}$$

to the half-plane $\{z \ : \ \mathrm{Re}\, z > 0\}$.

12
Final Remarks

Introduction

The purpose of this very brief final chapter is to make the point that complex analysis is a living topic. The first section describes the Riemann Hypothesis, perhaps the most remarkable unsolved problem in mathematics. Because it requires a great deal of mathematical background even to understand the conjecture, it is not as famous as the Goldbach Conjecture (every even number greater than 2 is the sum of two prime numbers) or the Prime Pairs Conjecture (there are infinitely many pairs (p, q) of prime numbers with $q = p + 2$) but it is hugely more important than either of these, for a successful proof would have many, many consequences in analysis and number theory.

The second and final section deals with iteration of complex functions, a topic that has given rise to arguably the most powerful visual images of twentieth century mathematics, and has demonstrated that fractal sets, far from being an isolated curiosity, occur as answers to simple and natural questions in analysis.

12.1 Riemann's Zeta Function

It is well known that, for real values of s, the series

$$1 + \frac{1}{2^s} + \frac{1}{3^s} + \cdots$$

is convergent if and only if $s > 1$. If we allow $s = \sigma + i\tau$ to be complex, then

$$|n^s| = |n^\sigma||n^{i\tau}| = |n^\sigma||e^{i\tau \log n}| = |n^\sigma|,$$

and so the series is (absolutely) convergent if $\sigma > 1$. We define **Riemann's Zeta Function** ζ by

$$\zeta(s) = \sum_{n=1}^{\infty} n^{-s} \qquad (\text{Re } s > 1). \tag{12.1}$$

An immediate connection with number theory is revealed by the following theorem, due to Euler, in which $\mathbf{P}$ denotes the set $\{2, 3, 5, \ldots\}$ of all prime numbers.

Theorem 12.1

$$\frac{1}{\zeta(s)} = \prod_{p \in \mathbf{P}} \left(1 - \frac{1}{p^s}\right) \qquad (\text{Re } s > 1).$$

Proof

Observe first that

$$\zeta(s)\left(1 - \frac{1}{2^s}\right) = \left(1 + \frac{1}{2^s} + \frac{1}{3^s} + \cdots\right)\left(1 - \frac{1}{2^s}\right) = 1 + \frac{1}{3^s} + \frac{1}{5^s} + \cdots,$$

all terms $1/n^s$, where n is even, being omitted. Next,

$$\zeta(s)\left(1 - \frac{1}{2^s}\right)\left(1 - \frac{1}{3^s}\right) = 1 + \frac{1}{5^s} + \frac{1}{7^s} + \frac{1}{11^s} + \cdots,$$

where now we are leaving out $1/n^s$ for all multiples of 2 or 3. If p_k is the kth prime, we see that

$$\zeta(s)\left(1 - \frac{1}{2^s}\right)\left(1 - \frac{1}{3^s}\right)\cdots\left(1 - \frac{1}{p_k^s}\right) = 1 + \sum_{n \in D_k} \frac{1}{n^s},$$

where D_k is the set of natural numbers not divisible by any of the primes $2, 3, \ldots, p_k$. Hence

$$\left|\zeta(s)\left(1 - \frac{1}{2^s}\right)\left(1 - \frac{1}{3^s}\right)\cdots\left(1 - \frac{1}{p_k^s}\right) - 1\right| \leq \left|\frac{1}{(p_k + 1)^s}\right| + \left|\frac{1}{(p_k + 2)^s}\right| + \cdots,$$

and this tends to 0 as $k \to \infty$. Hence

$$\zeta(s) \prod_{p \in \mathbf{P}} \left(1 - \frac{1}{p^s}\right) = 1 \,,$$

as required.   □

We have already (see Remark 9.17) come across the gamma function

$$\Gamma(s) = \int_0^\infty x^{s-1} e^{-x} \, dx \qquad (s > 0) \,, \tag{12.2}$$

and here too we can allow s to be complex and regard the function as defined whenever $\operatorname{Re} s > 0$. It is easily proved that

$$\Gamma(s) = (s-1)\Gamma(s-1) \qquad (\operatorname{Re} s > 1) \,, \tag{12.3}$$

and we can use this functional equation backwards to *define* $\Gamma(s)$ for $\operatorname{Re} s < 0$: if $\operatorname{Re}(s+n) \in (0,1)$, then

$$\Gamma(s) = \frac{\Gamma(s+n)}{s(s+1)\dots(s+n-1)} \,.$$

This fails if s is 0 or a negative integer, and in fact it can be shown that Γ is a meromorphic function with simple poles at $0, -1, -2 \dots$.

Substituting $x = nu$ in the integral (12.2) gives

$$n^{-s}\Gamma(s) = \int_0^\infty e^{-nu} u^{s-1} \, du \,,$$

and summing from 1 to ∞ gives

$$\zeta(s)\Gamma(s) = \sum_{n=1}^\infty \left[\int_0^\infty e^{-nu} u^{s-1} \, du\right] = \int_0^\infty (e^{-u} + e^{-2u} + \cdots) u^{s-1} \, du$$
$$= \int_0^\infty \frac{e^{-u} u^{s-1} \, du}{1 - e^{-u}} \,.$$

(The change in the order of integration and summation can be justified, but I am deliberately omitting formal details in this chapter.) It follows that

$$\zeta(s) = \frac{1}{\Gamma(s)} \int_0^\infty \frac{u^{s-1}}{e^u - 1} \, du \qquad (\operatorname{Re} s > 1) \,.$$

A more difficult formula, which I shall not prove (see [14]), gives

$$\zeta(s) = \frac{i\Gamma(1-s)}{2\pi} \int_C \frac{(-z)^{s-1}}{e^z - 1} \, dz \,, \tag{12.4}$$

where C is a (limiting) contour beginning and ending at $+\infty$ on the x-axis, encircling the origin once in a positive direction, but slender enough to exclude the poles $\pm 2i\pi, \pm 4i\pi, \ldots$ of the integrand.

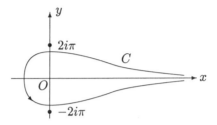

We interpret $(-z)^{s-1}$ in the usual way as $e^{(s-1)\log(-z)}$, noting that the cut for $\log(-z)$ lies along the positive x-axis.

The formula (12.4) makes sense for all s in $\mathbb{C}$, except possibly for the poles $2, 3, 4, \ldots$ of $\Gamma(1-s)$, but we already know that $\zeta(s)$ is defined at these points. In fact we now have $\zeta(s)$ defined as a meromorphic function over the whole of $\mathbb{C}$, with a single simple pole at $s = 1$.

By developing these ideas a little further (again see [14]) one obtains a functional equation for ζ, somewhat more complicated than Equation (12.3) for the gamma-function:

$$\zeta(s) = 2^{s-1}[\Gamma(s)]^{-1}\sec(\pi s/2)\zeta(1-s).$$

At each negative integer $[\Gamma(s)]^{-1}$ has a zero of order 1. If the integer is odd, then this is cancelled by the pole of order 1 for $\sec(\pi s/2)$, but if $s = -2, -4, \ldots$ we have $\zeta(s) = 0$. In fact those are the only zeros of ζ in the region $\{s : \operatorname{Re} s < 0\}$.

From (12.1) it is not hard to deduce that there are no zeros of ζ in the region $\{s : \operatorname{Re} s > 1\}$, and so we have the conclusion that the remaining zeros of ζ lie in the strip $\{s : 0 \le \operatorname{Re} s \le 1\}$. Riemann conjectured:

All the zeros of ζ in the strip $\{s : 0 \le \operatorname{Re} s \le 1\}$ lie on the line $\{s : \operatorname{Re} s = \frac{1}{2}\}$,

and this has become known as the **Riemann Hypothesis**.

It is something of a puzzle that (at the time of writing) this is still unproved, for complex analysis is replete with powerful results and techniques (a few of which appear in this book). The late twentieth century saw the solution of several of the classical unsolved problems, notably the Four Colour Theorem and the Fermat Theorem, but the Riemann Hypothesis has so far resisted all attempts. As early as 1914 Hardy[1] [8] proved that ζ has infinitely many zeros on the line $\operatorname{Re} s = \frac{1}{2}$, and nobody seriously believes that Riemann's guess is incorrect.

---

[1] Godfrey Harold Hardy, 1877–1947.

A much weaker version of the Hypothesis is that there are no zeros of ζ on the line $\mathrm{Re}\, s = 1$, and it was by proving this result that Hadamard[2] and de la Vallée Poussin[3] were able to establish the Prime Number Theorem: if $\pi(x)$ is defined as $|\{p \in \mathbf{P} : p \leq x\}|$, then

$$\pi(x) \sim \frac{x}{\log x}.$$

A precise error term in this formula would follow from the full Riemann Hypothesis.

There is an extensive literature on *consequences* of the Riemann Hypothesis, which is not as silly as it might seem at first sight. Titchmarsh[4], in his book *The zeta-function of Riemann* [15], at the beginning of the final "Consequences" chapter, puts the case very well:

> If the Riemann Hypothesis is true, it will presumably be proved some day. These theorems will then take their place as an essential part of the theory. If it is false, we may perhaps hope in this way sooner or later to arrive at a contradiction. Actually the theory, as far as it goes, is perfectly coherent, and shews no sign of breaking down.

As the spelling "shews" might suggest, Titchmarsh was writing in 1930, but his summary is just as true in 2003.

The classic texts by Titchmarsh [14, 15] and Whittaker and Watson [16] are an excellent source of further information.

12.2 Complex Iteration

The first hint that simple and natural questions in complex analysis might have unexpectedly complicated answers came in a question posed by Cayley[5] in 1879. Let us first remind ourselves of Newton's[6] method for finding approximate solutions to equations. Let f be a real function. If x_0 is chosen appropriately and if, for all $n \geq 0$,

$$x_{n+1} = x_n - \frac{f(x_n)}{f'(x_n)}, \tag{12.5}$$

then the sequence (x_n) tends to a root of the equation $f(x) = 0$. The term "appropriately" is deliberately vague, for $f(x) = 0$ may have several roots,

---

[2] Jacques Salomon Hadamard, 1865–1963.
[3] Charles Jean Gustave Nicolas Baron de la Vallée Poussin, 1866–1962.
[4] Edward Charles Titchmarsh, 1899–1963.
[5] Arthur Cayley, 1821–1895.
[6] Isaac Newton, 1643–1727.

and an inappropriate choice may well lead to a divergent sequence (x_n). For example, if $f(x) = x/(x^2 + 1)$, then the only root of $f(x) = 0$ is 0, but any choice of x_0 for which $|x_0| > 1$ leads to a divergent sequence (x_n).

The formula (12.5) makes sense if we interpret it for a complex function: we rewrite it (for psychological rather than logical reasons) as

$$z_{n+1} = z_n - \frac{f(z_n)}{f'(z_n)}, \qquad (12.6)$$

where z_0 is an arbitrary starting point. If $f(z) = 0$ has roots $\alpha_1, \alpha_2, \ldots, \alpha_m$, then there are **basins of attraction** $B_1, B_2, \ldots, B_m$ in the complex plane defined by

$$B_i = \{z_0 \in \mathbb{C} : \lim_{n \to \infty} z_n = \alpha_i\} \quad (i = 1, 2, \ldots, m).$$

For a linear function $z - k$ there is just one basin, namely $\mathbb{C}$ itself, and for the quadratic function $z^2 - 1$, with two zeros 1 and -1, there are two basins

$$B_1 = \{z \in \mathbb{C} : \operatorname{Re} z > 0\} \quad \text{and} \quad B_{-1} = \{z \in \mathbb{C} : \operatorname{Re} z < 0\}.$$

This much is straightforward and unsurprising. When it came to the cubic function $z^3 - 1$, Cayley remarked that "it appears to present considerable difficulty". It does indeed, for the basin of attraction of the root 1 looks like this:

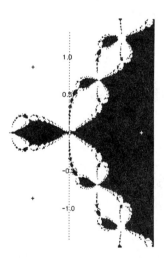

Cayley, of course, had no access to a computer, and could not possibly have guessed that the answer would be so complicated.

The process described by (12.6) can be defined in a different way. Given the function f we can define

$$F(z) = z - \frac{f(z)}{f'(z)}$$

and write
$$z_1 = F(z_0), \; z_2 = F(F(z_0)), \ldots.$$

The functions $F \circ F$, $F \circ F \circ F$, ... are written F^2, F^3, ..., and are called **iterates** of F. The basin of attraction of a root α of $f(z) = 0$ is then the set

$$B_\alpha = \{ z \in \mathbb{C} \; : \; \lim_{n \to \infty} F^n(z) = \alpha \}.$$

It is this process of **iteration** that gives rise to Julia[7] sets.

Let g be a polynomial function, and, for $n = 1, 2, \ldots$, let $g^n = \overbrace{g \circ g \circ \cdots \circ g}^{n}$ be the nth iterate of g. The **filled in Julia set** F of g is defined by

$$F = \{ z \in \mathbb{C} \; : \; g^n(z) \not\to \infty \text{ as } n \to \infty \},$$

and the boundary ∂F of F is called the **Julia set** of g.

Consider the simplest possible quadratic function $g : z \mapsto z^2$, where $g^n(z) = z^{2^n}$. Here it is clear that

$$F = \{ z \; : \; |z| \leq 1 \}, \quad \partial F = \{ z \; : \; |z| = 1 \},$$

an unexciting conclusion, but it makes the point that not all Julia sets are "funny". The situation changes dramatically if we consider the quadratic function $f_c : z \mapsto z^2 + c$, with $c \neq 0$. If $c = (0.6)i$ the filled in Julia set looks like this:

---

[7] Gaston Maurice Julia, 1893–1978.

If $c = -0.2 + (0.75)i$ it looks like this:

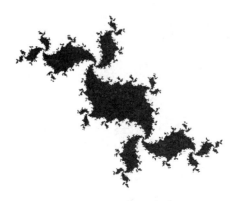

The **Mandelbrot**[8] set, depicted below, is defined as

$$M = \{c \in \mathbb{C} \, : \, f_c^{\,n}(0) \not\to \infty\}\,.$$

It seems appropriate to end this book with a picture that is the most striking of all the images of late twentieth century mathematics. This has been a
flimsy account of an important area. Much solid and fascinating mathematics
is involved in a proper study, but this is well beyond the scope of an introduc-
tory book. For a mathematical account of fractal sets, including Julia sets and
the Mandelbrot set, see [7]. For a more visual account, with lots of excellent
pictures, see [12].

---

[8] Benoit Mandelbrot, 1924–.

Chapter 1

1.1. Let A be bounded below by b. Then $-A = \{x \in \mathbb{R} : -x \in A\}$ is bounded above by $-b$, and so has a least upper bound c. Then $-c$ is the greatest lower bound of A.

1.2. The result certainly holds for $n = 1$. Suppose that it holds for $n - 1$. Then $q_n = (3/2)(3^{n-1} + 1) - 1 = (1/2)(3^n + 3 - 2) = (1/2)(3^n + 1)$.

1.3. It is useful to note first that γ and δ are the roots of the equation $x^2 - x - 1 = 0$. Thus

$$\gamma^2 = \gamma + 1, \quad \delta^2 = \delta + 1. \tag{13.1}$$

Since $(1/\sqrt{5})(\gamma - \delta) = 1$ and $(1/\sqrt{5})(\gamma^2 - \delta^2) = (1/\sqrt{5})(\gamma + 1 - \delta - 1) = 1$, the result holds for $n = 1$ and $n = 2$. Let $n \geq 3$ and suppose that the result holds for all $k < n$. Then $f_n = (1/\sqrt{5})[\gamma^{n-1} - \delta^{n-1} + \gamma^{n-2} - \delta^{n-2}] = (1/\sqrt{5})[\gamma^{n-2}(\gamma + 1) - \delta^{n-2}(\delta + 1)] = (1/\sqrt{5})[\gamma^n - \delta^n]$, by (13.1).

1.4. From elementary trigonometry we have $2\sin(\pi/4)\cos(\pi/4) = \sin(\pi/2) = 1$. Also, $\sin((\pi/2) - \theta) = \sin(\pi/2)\cos(-\theta) + \cos(\pi/2)\sin(-\theta) = \cos\theta$, and so $\sin(\pi/4) = \cos(\pi/4)$. Thus $2\cos^2(\pi/4) = 1$, and so $\cos(\pi/4)$, being positive, is equal to $1/\sqrt{2}$.

1.5. a) We easily deduce that $\cos 2\theta = 2\cos^2\theta - 1$ and $\sin 2\theta = 2\sin\theta\cos\theta$. Then $\cos 3\theta = \cos 2\theta \cos\theta - \sin 2\theta \sin\theta = (2\cos^2\theta - 1)\cos\theta - 2\cos\theta(1 - \cos^2\theta) = 4\cos^3\theta - 3\cos\theta$. Put $\theta = \pi/6$; then $4\cos^3(\pi/6) - 3\cos(\pi/6) =$

$\cos(\pi/2) = 0$, and so $\cos(\pi/6)$, being positive, is equal to $\sqrt{3}/2$. It follows that $\sin(\pi/6) = \sqrt{1 - \cos^2(\pi/6)} = 1/2$.

b) These follow immediately from $\sin((\pi/2) - \theta) = \cos\theta$.

1.6. We know that $\cos(u + v) = \cos u \cos v - \sin u \sin v$ and easily deduce that $\cos(u - v) = \cos u \cos v + \sin u \sin v$. Hence $\cos(u + v) + \cos(u - v) = 2\cos u \cos v$ and $\cos(u+v) - \cos(u-v) = 2\sin u \sin(-v)$. The result follows if we let $u = (x + y)/2$ and $v = (x - y)/2$.

1.7. We verify that $a_1 = 2^0 \cos 0 = 1$, $a_2 = 2^1 \cos(\pi/3) = 1$. Suppose that $n \geq 3$ and that $a_k = 2^{k-1} \cos((k - 1)\pi/3)$ for all $k < n$. Then

$$a_n = 2a_{n-1} - 4a_{n-2} = 2^{n-1}[\cos \tfrac{(n-2)\pi}{3} - \cos \tfrac{(n-3)\pi}{3}]$$

$$= 2^{n-1}[\cos \tfrac{(n-2)\pi}{3} + \cos \tfrac{n\pi}{3}] \text{ (since } \cos(\theta - \pi) = -\cos\theta)$$

$$= 2^n \cos \tfrac{(n-1)\pi}{3} \cos \tfrac{\pi}{3} = 2^{n-1} \cos \tfrac{(n-1)\pi}{3} .$$

Chapter 2

2.1. a) $M(a,b)M(c,d) = M(ac - bd, ad + bc)$.
     b) This is routine.
     c) $M(0,1)M(0,1) = M(0.0 - 1.1, 0.1 - 1.0) = M(-1,0)$.
     d)

$$M(a,b) = \begin{pmatrix} a & b \\ -b & a \end{pmatrix} = \begin{pmatrix} a & 0 \\ a & 0 \end{pmatrix} + b \begin{pmatrix} 0 & 1 \\ -1 & 0 \end{pmatrix} = a + bi .$$

2.2. $x = 1 \pm \sqrt{-4} = 1 \pm 2i$.

2.3. Let $z = x + iy$. Then $iz = ix - y$ and so $\operatorname{Re}(iz) = -y = -\operatorname{Im} z$, $\operatorname{Im}(iz) = x = \operatorname{Re} z$.

2.4. a) $(3 + 2i)/(1 + i) = [(3 + 2i)(1 - i)]/2 = \tfrac{1}{2}(5 - i)$.
     b) $(1 + i)/(3 - i) = [(1 + i)(3 + i)]/10 = \tfrac{1}{10}(2 + 4i) = \tfrac{1}{5}(1 + 2i)$.
     c) $(z + 2)(\bar{z} + 1) = [(z + 2)(\bar{z} + 1)]/[(z + 1)(\bar{z} + 1)] = [(x^2 + y^2 + 3x + 2) - yi]/(x^2 + y^2 + 2x + 1)$.

2.5. a) $|z| = \sqrt{2}$, $\arg z = -\pi/4$;   b) $|z| = 3$, $\arg z = -\pi/2$:   c) $|z| = 5$, $\arg z = \tan^{-1}(4/3)$;   d) $|z| = \sqrt{5}$; $\arg z = \pi + \tan^{-1}(-2)$.

2.6. Let $c = re^{i\theta}$, $d = \rho e^{i\phi}$, with $\theta = \arg c$, $\phi = \arg d$. Then $c/d = (r/\rho)e^{i(\theta - \phi)}$. So, taking account of the potential ambiguity in arg, we have $\arg(c/d) \equiv \theta - \phi$.

2.7. $1 + i = \sqrt{2}e^{i\pi/4}$; so $(1 + i)^{16} = 2^8 e^{4i\pi} = 2^8 = 256$.

2.8. $2 + 2i\sqrt{3} = 4e^{i\pi/3}$; so $(2 + 2i\sqrt{3})^9 = 2^{18}e^{3i\pi} = -2^{18}$.

2.9. $i^4 = 1$; so $i^{4q} = 1$ for every q in $\mathbb{Z}$. Thus $i^{4q+1} = i$, $i^{4q+2} = i^2 = -1$, $i^{4q+3} = i^3 = -i$.

2.10. $\sum_{n=0}^{100} i^n = (1 - i^{101})/(1 - i) = (1 - i)/(1 - i) = 1$.

2.11. When $n = 1$ the right hand side is $(1 - 2z + z^2)/(1 - z)^2 = 1$, and so the result is true when $n = 1$. Suppose that it holds for $n = k$. Then

$$1 + 2z + \cdots (k+1)z^k = \frac{1 - (k+1)z^k + kz^{k+1}}{(1-z)^2} + (k+1)z^k$$

$$= \frac{1 - (k+1)z^k + kz^{k+1} + (k+1)z^k - 2(k+1)z^{k+1} + (k+1)z^{k+2}}{(1-z^2)}$$

$$= \frac{1 - (k+2)z^{k+1} + (k+1)z^{k+2}}{(1-z)^2},$$

and so the result holds by induction for all n. The sum to infinity follows easily.

2.12. If a is positive and $0 < r < 1$, then $a + ar + \cdots ar^{n-1} < a/(1-r)$. Put $a = 1$ and $r = |z_2/z_1|$ to obtain $1 + |z_2/z_1| + \cdots + |z_2/z_1|^{n-1} < 1/[1 - |z_2/z_1|]$. The left hand side is greater than n times its smallest term, so greater than $n|z_2/z_1|^{n-1}$. Hence $n|z_2/z_1|^{n-1} < 1/[1 - |z_2/z_1|] = |z_1|/(|z_1| - |z_2|)$.

2.13. $|z_1 + z_2|^2 + |z_1 - z_2|^2 = z_1\bar{z}_1 + z_1\bar{z}_2 + z_2\bar{z}_1 + z_2\bar{z}_2 + z_1\bar{z}_1 - z_1\bar{z}_2 - z_2\bar{z}_1 + z_2\bar{z}_2 = 2(z_1\bar{z}_1 + z_2\bar{z}_2) = 2(|z_1|^2 + |z_2|^2)$. Hence, putting $z_1 = c$, $z_2 = \sqrt{c^2 - d^2}$, we have $\left[|c + \sqrt{c^2 - d^2}| + |c - \sqrt{c^2 - d^2}|\right]^2 = \left[|z_1 + z_2| + |z_1 - z_2|\right]^2 = |z_1 + z_2|^2 + |z_1 - z_2|^2 + 2|(z_1 + z_2)(z_1 - z_2)| = 2(|z_1|^2 + |z_2|^2) + 2|z_1^2 - z_2^2| = 2|c|^2 + 2|c^2 - d^2| + 2|d|^2 = |c + d|^2 + |c - d|^2 + 2|c + d||c - d| = \left[|c + d| + |c - d|\right]^2$. The result follows if we take square roots.

2.14. $e^{i\theta} + e^{3i\theta} + \cdots + e^{(2n+1)i\theta} = e^{i\theta}[e^{(2n+2)i\theta} - 1]/[e^{2i\theta} - 1] = [e^{(2n+2)i\theta} - 1]/[e^{i\theta} - e^{-i\theta}] = [e^{(2n+2)i\theta} - 1]/2i\sin\theta = (1/2\sin\theta)\left[-i\big(\cos(2n+2)\theta + i\sin(2n+2)\theta\big) + i\right]$, of which the real part is $[\sin(2n+2)\theta]/[2\sin\theta]$.

2.15. Since $a_n\gamma^n + a_{n-1}\gamma^{n-1} + \cdots + a_1\gamma + a_0 = 0$, the complex conjugate is also equal to 0; that is, since the coefficients a_i are all real, $a_n\bar{\gamma}^n + a_{n-1}\bar{\gamma}^{n-1} + \cdots + a_1\bar{\gamma} + a_0 = 0$. Thus $\gamma = \rho e^{i\theta}$ and $\bar{\gamma} = \rho e^{-i\theta}$ are both roots and so $P(z)$ is divisible by $(z - \rho e^{i\theta})(z - \rho e^{-i\theta}) = z^2 - 2\rho\cos\theta + \rho^2$.

2.16. a) By the standard formula, $z = \frac{1}{2}\left(3 - i \pm \sqrt{(3 - i)^2 - 4(4 - 3i)}\right) = \frac{1}{2}\left(3 - i \pm \sqrt{-8 + 6i}\right) = \frac{1}{2}\left(3 - i \pm (1 + 3i)\right) = 2 + i$ or $1 - 2i$.

    b) Similarly, $z = \frac{1}{2}[(3 + i) \pm \sqrt{(3 + i)^2 - 8 - 4i}] = \frac{1}{2}\left(3 + i \pm \sqrt{2i}\right) = \frac{1}{2}[3 + i \pm (1 + i)] = 2 + i$ or 1.

2.17. a) $\{z : |2z + 3| \le 1\}$ is the circular disc with centre $-\frac{3}{2}$ and radius $\frac{1}{2}$.

b) $|z|^2 \geq |2z+1|^2$ if and only if $x^2 + y^2 \geq (2x+1)^2 + (2y)^2$, that is, if and only if $3x^2 + 3y^2 + 4x + 1 \geq 0$, that is, if and only if $\left(x + \frac{2}{3}\right)^2 + y^2 \geq \frac{1}{9}$. Thus the set is the exterior of the circular disc $\{z : |z + \frac{2}{3}| < \frac{1}{3}\}$.

2.18. The roots of $z^5 = 1$ are 1, $e^{\pm 2i\pi/5}$, $e^{\pm 4i\pi/5}$, and so $z^5 - 1 = (z-1)[z^2 - 2z\cos(2\pi/5) + 1][z^2 - 2z\cos(4\pi/5) + 1]$. Since we also have $z^5 - 1 = (z-1)(z^4 + z^3 + z^2 + z + 1)$, we deduce that $[z^2 - 2z\cos(2\pi/5) + 1][z^2 - 2z\cos(4\pi/5) + 1] = z^4 + z^3 + z^2 + z + 1$. Equating coefficients of z^3 gives $\cos(2\pi/5) + \cos(4\pi/5) = -\frac{1}{2}$, and equating coefficients of z^2 gives $\cos(2\pi/5)\cos(4\pi/5) = -\frac{1}{4}$. Hence $\cos(2\pi/5)$ and $\cos(4\pi/5)$ are the roots of the equation $x^2 + \frac{1}{2}x - \frac{1}{4} = 0$. The roots of this equation are $\frac{1}{4}(-1 \pm \sqrt{5})$ and, since $\cos(2\pi/5)$ is positive and $\cos(4\pi/5)$ is negative, we must have $\cos(2\pi/5) = \frac{1}{4}(\sqrt{5} - 1)$, $\cos(\pi/5) = -\cos(4\pi/5) = \frac{1}{4}(\sqrt{5} + 1)$.

Chapter 3

3.1. Let c be a point in $\mathbb{C}$ not lying in the real interval $[a, b]$. Let $\delta = \min\{|c - z| : z \in [a, b]\}$. Then $\delta > 0$, and the neighbourhood $N(c, \delta/2)$ lies wholly outside $[a, b]$. Thus $[a, b]$ is closed. On the other hand, (a, b) is not open, since for every c in (a, b) there is no neighbourhood $N(c, \delta)$ of c lying wholly inside (a, b). It is not closed either, for $a \notin (a, b)$, yet every neighbourhood $N(a, \delta)$ of a intersects (a, b).

3.2. Let $z \in A$; thus $|z| = r$, where $1 < r < 2$. If $\delta = \min\{r - 1, 2 - r\}$, then $N(z, \delta) \subseteq A$. Hence A is open. The closure of A is $\{z : 1 \leq z \leq 2\}$, and $\partial A = \kappa(0, 1) \cup \kappa(0, 2)$.

3.3. If $f = g + h$, with $|g(z)| \leq K|z|^2$, $|h(z)| \leq L|z|^3$ for all sufficiently small z, it follows (since $|z|^3 \leq |z|^2$ for all $|z| \leq 1$), that $|f(z)| \leq |g(z)| + |h(z)| \leq (K + L)|z|^2$ for all sufficiently small z. Thus $O(z^2) + O(z^3) = O(z^2)$ as $z \to 0$.

If $f = g + h$, with $|g(z)| \leq K|z|^2$, $|h(z)| \leq L|z|^3$ for all sufficiently *large* z, it follows (since $|z|^3 > |z|^2$ for all $|z| > 1$), that $|f(z)| \leq |g(z)| + |h(z)| \leq (K + L)|z|^3$ for all sufficiently large z. Thus $O(z^2) + O(z^3) = O(z^3)$ as $z \to \infty$.

3.4.
$$\frac{1}{z}[(1 + z)^n - (1 + nz)] = \sum_{r=2}^{n} \binom{n}{r} z^{r-1} \to 0 \text{ as } z \to 0.$$

3.5.

$$\frac{3z^2 + 7z + 5}{(z+1)^2} = \left(3 + \frac{7}{z} + \frac{5}{z^2}\right)\left(1 + \frac{1}{z^2}\right)^{-2}$$

$$= \left(3 + \frac{7}{z} + \frac{5}{z^2}\right)\left(1 - \frac{2}{z} + O(z^{-2})\right) \quad \text{(by Example 3.11)}$$

$$= 3 + \frac{1}{z} + O(z^{-2})$$

as $z \to \infty$.

3.6. For all z such that $|z| \le 1$, $|p(z)| \le |a_0| + |a_1| + \cdots + |a_n|$. So $p(z) = O(1)$ as $z \to 0$. Also, for all $|z| \ge 1$

$$|p(z)| = |z|^n \left| \left(a_n + \frac{a_{n-1}}{z} + \cdots + \frac{a_0}{z^n}\right)\right| \le |z|^n (|a_n| + |a_{n-1}| + \cdots + |a_0|),$$

and so $p(z) = O(z^n)$ as $n \to \infty$.

Chapter 4

4.1.   a) $f(z) = i(x^2 - y^2 + 2ixy) + 2(x + iy) = (2x - 2xy) + i(x^2 - y^2 + 2y)$; so $u = 2x - 2xy$, $v = x^2 - y^2 + 2y$. Thus $\partial u/\partial x = \partial u/\partial y = 2 - 2y$, $\partial v/\partial x = -\partial u/\partial y = 2x$.

   b) Multiply numerator and denominator by the conjugate of the denominator to make the denominator real: $f(z) = (z + i)(2\bar{z} + 3i)/(2z - 3i)(2\bar{z}+3i) = (2z\bar{z}+3iz+2i\bar{z}-3)/(4z\bar{z}+6i(z-\bar{z})+9) = \left(2(x^2+y^2)+ 3i(x+iy) + 2i(x-iy) - 3\right)/\left(4(x^2 - y^2) - 12y + 9\right)$. Hence $u = (2x^2 + 2y^2 - y - 3)/(4x^2 + 4y^2 - 12y + 9)$, $v = 5x/(4x^2 + 4y^2 - 12y + 9)$. Then verify that $\partial u/\partial x = \partial v/\partial y = (60x - 40xy)/(4x^2 + 4y^2 - 12y + 9)^2$, $\partial v/\partial x = -\partial u/\partial y = (20y^2 - 20x^2 - 60y + 45)/(4x^2 + 4y^2 - 12y + 9)^2$.

4.2.   Since $a^2 + b^2 < R^2$ and $c^2 + d^2 < R^2$, it follows that $(a^2 + d^2) + (b^2 + c^2) < 2R^2$. Hence at least one of $a^2 + d^2$ and $b^2 + c^2$ is less than R^2.

4.3.   Suppose first that f is differentiable at c. For all $z \ne c$, let $A(z) = \left(f(z) - f(c)\right)/(z - c)$. Certainly $f(z) = f(c) + A(z)(z - c)$. Then $\lim_{z \to c} A(z) = f'(c)$, and so A is continuous if we define $A(c) = f'(c)$.
Conversely, suppose that A exists, and is continuous at c. Then, for all $z \ne c$,

$$A(z) = \frac{f(z) - f(c)}{z - c}.$$

Since A is continuous at c, the limit

$$\lim_{z \to c} \frac{f(z) - f(c)}{z - c}$$

exists. Thus f is differentiable at c.

4.4.
$$\frac{p(z)}{a_n z^n} = 1 + \frac{a_{n-1}}{a_n z} + \cdots + \frac{a_0}{a_n z^n} \to 1 \text{ as } |z| \to \infty.$$

4.5. $\overline{e^z} = \overline{e^x e^{iy}} = e^x e^{-iy} = e^{\bar z}$. Hence
$$\overline{\sin z} = \overline{(1/2i)(e^{iz} - e^{-iz})} = (-1/2i)(e^{-i\bar z} - e^{i\bar z}) = \sin \bar z,$$

and
$$\overline{\cos z} = \overline{(1/2)(e^{iz} + e^{-iz})} = (1/2)(e^{-i\bar z} + e^{i\bar z}) = \cos \bar z.$$

4.6. $\cosh z \cosh w + \sinh z \sinh w = \frac{1}{4}[(e^z + e^{-z})(e^w + e^{-w}) + (e^z - e^{-z})(e^w - e^{-w})] = \frac{1}{4}[e^{z+w} + e^{-z+w} + e^{z-w} + e^{-z-w} + e^{z+w} - e^{-z+w} - e^{z-w} + e^{-z-w}] = \frac{1}{2}(e^{z+w} + e^{-(z+w)}) = \cosh(z+w)$. Similarly $\sinh z \cosh w + \cosh z \sinh w = \frac{1}{4}[(e^z - e^{-z})(e^w + e^{-w}) + (e^z + e^{-z})(e^w - e^{-w})] = \frac{1}{4}[e^{z+w} - e^{-z+w} + e^{z-w} - e^{-z-w} + e^{z+w} + e^{-z+w} - e^{z-w} - e^{-z-w}] = \frac{1}{2}(e^{z+w} - e^{-(z+w)}) = \sinh(z+w)$.

4.7. $F'(z) = 2\cosh z \sinh z - 2\sinh z \cosh z = 0$ for all z. Hence, by Theorem 4.9, $F(z)$ is constant throughout $\mathbb{C}$. Since $F(0) = 1$, $F(z) = 1$ for all z.

4.8. $\cos(iz) = \frac{1}{2}(e^{i(iz)} + e^{-i(iz)}) = \frac{1}{2}(e^{-z} + e^z) = \cosh z$; $\sin(iz) = \frac{1}{2i}(e^{i(iz)} - e^{-i(iz)}) = (-i)\frac{1}{2}(e^{-z} - e^z) = \frac{i}{2}(e^z - e^{-z}) = i\sinh z$. Hence $\cos z = \cos(x + iy) = \cos x \cos(iy) - \sin x \sin(iy) = \cos x \cosh y - i\sin x \sinh y$. So $u = \cos x \cosh y$, $v = -\sin x \sinh y$, and $\partial u/\partial x = \partial v/\partial y = -\sin x \cosh y$, $\partial v/\partial x = -\partial u/\partial y = -\cos x \sinh y$.
Similarly, $\sin z = \sin(x + iy) = \sin x \cos iy + \cos x \sin iy = \sin x \cosh y + i\cos x \sinh y$. Thus $u = \sin x \cosh y$, $v = \cos x \sinh y$, and $\partial u/\partial x = \partial v/\partial y = \cos x \cosh y$, $\partial v/\partial x = -\partial u/\partial y = -\sin x \sinh y$.
For the second part, use the identities $\sin^2 x + \cos^2 x = 1$ and $\cosh^2 x - \sinh^2 x = 1$ to show that $|\sin z|^2 = \sin^2 x \cosh^2 y + \cos^2 x \sinh^2 y = \sin^2 x(\sinh^2 y + 1) + (1 - \sin^2 x)\sinh^2 y = \sin^2 x + \sinh^2 y$. Similarly, $|\cos z|^2 = \cos^2 x(1 + \sinh^2 y) + (1 - \cos^2 x)\sinh^2 y = \cos^2 x + \sinh^2 y$.

4.9. Since $\cos(iy) = \cosh y$ and $\sin(iy) = i\sinh y$, we have that $|\cos(iy)| = \frac{1}{2}(e^y + e^{-y}) > \frac{1}{2}e^y$. Also $|\sin(iy)| = \frac{1}{2}|e^y - e^{-y}|$. If $y \geq 0$ then $e^{-y} \leq 1 \leq e^y$, and so $|\sin(iy)| \geq \frac{1}{2}(e^y - 1)$; if $y \leq 0$ then $e^y \leq 1 \leq e^{-y}$, and so $|\sin(iy)| \geq \frac{1}{2}(e^{-y} - 1)$. Combining the two inequalities gives $|\sin(iy)| \geq \frac{1}{2}(e^{|y|} - 1)$ for all real y. Both cos and sin are unbounded in $\mathbb{C}$.

4.10. $\cos \pi = -1$, while $\sin \pi = 0$. Hence $\cos 2\pi = \cos^2 \pi - \sin^2 \pi = 1$, while $\sin 2\pi = 2\sin \pi \cos \pi = 0$. It now follows that $\sin(z + \pi) = \sin z \cos \pi + \cos z \sin \pi = -\sin z$ and $\cos(z + \pi) = \cos z \cos \pi - \sin z \sin \pi = -\cos z$. From these formulae it follows (and can be proved formally by induction) that $\sin(z + n\pi) = (-1)^n \sin z$ and $\cos(z + n\pi) = (-1)^n \cos z$.

4.11. $\cosh(z + 2\pi i) = \cos i(z + 2\pi i) = \cos(iz - 2\pi) = \cos iz = \cosh z$, $\sinh(z + 2\pi i) = -i\sin i(z + 2\pi i) = -i\sin(iz - 2\pi) = -i\sin iz = \sinh z$.

4.12. First, $e^{z^2} = e^{x^2 - y^2 + 2xyi} = e^{x^2 - y^2}e^{2xyi} = e^{x^2 - y^2}\big(\cos(2xy) + i\sin(2xy)\big)$. So $u = e^{x^2 - y^2}\cos(2xy)$, $v = e^{x^2 - y^2}\sin(2xy)$. As a check, observe that $\partial u/\partial x = \partial v/\partial y = e^{x^2 - y^2}\big(2x\cos(2xy) - 2y\sin(2xy)\big)$, $\partial v/\partial x = -\partial u/\partial y = e^{x^2 - y^2}\big(2x\sin(2xy) + 2y\cos(2xy)\big)$.
Next,

$$e^{e^z} = e^{e^x\cos y + ie^x\sin y} = \big(e^{e^x\cos y}\big)\big(e^{ie^x\sin y}\big)$$
$$= e^{e^x\cos y}\big(\cos(e^x\sin y) + i\sin(e^x\sin y)\big).$$

Thus $u = e^{e^x\cos y}\cos(e^x\sin y)$, $v = e^{e^x\cos y}\sin(e^x\sin y)$. The verification of the Cauchy–Riemann equations is a pleasant exercise in partial differentiation.

4.13. $|\sin(x + iy)| = \frac{1}{2}|e^{ix - y} - e^{-ix + y}| \geq \frac{1}{2}\big||e^{ix}e^{-y}| - |e^{-ix}e^y|\big| = \frac{1}{2}(e^y - e^{-y}) = \sinh y$.

4.14. Since the series converges, there exists $K > 0$ such that

$$\Big|\sum_{n=2}^{\infty}(1/n!)\Big| \leq K.$$

Hence, for all $|z| \leq 1$,

$$|e^z - 1 - z| \leq K|z|^2.$$

Again,

$$\frac{\cos z - \big(1 - (z^2/2)\big)}{z^3} = \frac{z}{4!} - \frac{z^3}{6!} + \cdots,$$

and this tends to 0 as $z \to 0$.

4.15. As a multifunction, $z^i = e^{i\operatorname{Log} z} = \{e^{i(\log|z| + i\arg z + 2n\pi i)} : n \in \mathbb{Z}\} = \{e^{i\log|z|}e^{-\arg z - 2n\pi} : n \in \mathbb{Z}\}$. For $z = -i$ we have $\log|z| = 0$ and $\arg z = -\pi/2$. So $(-i)^i = \{e^{(\pi/2) - 2n\pi} : n \in \mathbb{Z}\}$.

4.16. $w \in \operatorname{Sin}^{-1} z \iff \frac{1}{2i}(e^{iw} - e^{-iw}) = z \iff e^{2iw} - 2ize^{iw} - 1 = 0 \iff e^{iw} = iz \pm \sqrt{1 - z^2} \iff w \in -i\operatorname{Log}(iz \pm \sqrt{1 - z^2})$. If $z = 1/\sqrt{2}$ then $\operatorname{Sin}^{-1} z = -i\operatorname{Log}(\frac{i}{\sqrt{2}} \pm \frac{1}{\sqrt{2}}) = -i\big(\operatorname{Log}(e^{i\pi/4}) \cup \operatorname{Log}(e^{i(\pi - \pi/4)})\big) = \{n\pi + (-1)^n\pi/4 : n \in \mathbb{Z}\}$.

4.17. $w \in \operatorname{Tan}^{-1} z \iff \sin w/\cos w = z \iff (e^{2iw} - 1)/(e^{2iw} + 1) = iz \iff e^{2iw}(1 - iz) = 1 + iz \iff w \in (1/2i)\operatorname{Log}\big(1 + iz)/(1 - iz)\big)$.
Putting $z = e^{i\theta}$ gives $(1 + iz)/(1 - iz) = [(1 + iz)(1 + i\bar{z})]/[(1 - iz)(1 + i\bar{z})] = (1 + 2i\operatorname{Re} z - z\bar{z})/(1 + 2\operatorname{Im} z + z\bar{z}) = i\cos\theta/(1 + \sin\theta)$, a complex number

with argument $\pi/2$ (since $\cos\theta/(1+\sin\theta)$ is positive in $(-\pi/2,\pi/2)$). Hence $\mathrm{Tan}^{-1}(e^{i\theta}) = \{(1/2i)\left[\log\left(\cos\theta/(1+\sin\theta)\right)+(2n+\tfrac{1}{2})\pi i\right] : n \in \mathbb{Z}\}$, of which the real part is $\{n+\tfrac{1}{4}\pi : n \in \mathbb{Z}\}$.

4.18. $(-1)^{-i} = \exp\left(-i\,\mathrm{Log}(-1)\right)$, of which the principal value is

$$\exp\left((-i)(i\pi)\right) = \exp(\pi)\,.$$

The logarithm of this is π.

4.19. The function $\sin z/\cos z$ has singularities where $\cos z = 0$, that is, at the points $(2n+1)\pi/2$. Now $\cos\left(w + (2n+1)\pi/2\right) = \cos w\cos(2n+1)\pi/2 - \sin w\sin(2n+1)\pi/2 = (-1)^{n+1}\sin w$, and $\sin\left(w+(2n+1)\pi/2\right) = \sin w\cos(2n+1)\pi/2 + \cos w\sin(2n+1)\pi/2 = (-1)^n\cos w$. Hence, putting $w = z - (2n+1)\pi/2$, we see that $\lim_{z\to(2n+1)\pi/2}(z-(2n+1)\pi/2)\tan z = \lim_{w\to 0}w(-\cos w/\sin w) = -1$. So the singularities are all simple poles.

4.20. $\sin z = 0$ if and only if $z = n\pi$. Now, $\sin(w+n\pi) = (-1)^n\sin w$, and so, if $n \neq 0$, $\lim_{z\to n\pi}(z-n\pi)(1/z\sin z) = \lim_{w\to 0}\left(w/((w+n\pi)(-1)^n\sin w)\right) = (-1)^n/n\pi$. If $n = 0$, then $\lim_{z\to 0}z^2(1/z\sin z) = 1$. There are simple poles at $z = n\pi$ $(n = \pm 1, \pm 2, \ldots)$, and a double pole at $z = 0$.

4.21. Let $r(z) = p(z)/(z-c)^k q(z)$, where $p(c)$ and $q(c)$ are non-zero. Then $r'(z) = \left[(z-c)^k q(z)p'(z) - \left((z-c)^k q'(z) + k(z-c)^{k-1}q(z)\right)p(z)\right]/\left[(z-c)^k q(z)\right]^2 = \left[(z-c)\left(q(z)p'z) - q'(z)p(z)\right) - kq(z)p(z)\right]/(z-c)^{k+1}\left(q(z)\right)^2$. Then $(z-c)^k r'(z) =\to \infty$ as $z \to c$, and $\lim_{z\to c}(z-c)^{k+1}r'(z) = -kp(c)/q(c)$. Thus c is a pole of order $k+1$.

Chapter 5

5.1. Let S be the open disc $N(0,1)$, so that S is bounded but not closed. If $f(z) = 1/(1-z)$, then f is continuous but not bounded in S. Next, let $S = \mathbb{C}$, so that f is closed but not bounded. If $f(z) = z$, then f is continuous but not bounded.

5.2. The length is $\int_0^1 |\gamma'(t)|\,dt = \int_0^1 |d-c|\,dt = |d-c|$.

5.3.   a) The curve is an ellipse:

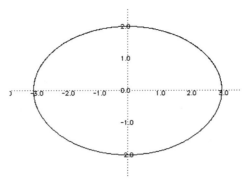

b) The curve is part of a hyperbola:

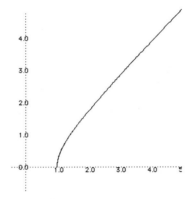

c) The curve is one branch of a rectangular hyperbola:

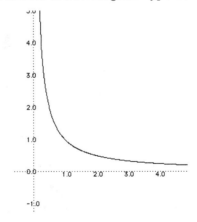

5.4. The curve is a spiral:

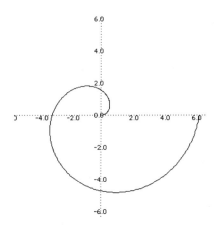

Writing $\gamma(t) = te^{it}$, we have $\gamma'(t) = (1 + it)e^{it}$ and so $|\gamma'(t)| = \sqrt{1 + t^2}$. Hence

$$\Lambda(\gamma^*) = \int_0^{2\pi} \sqrt{1 + t^2} \, dt = \left[\tfrac{1}{2}t\sqrt{1 + t^2} + \tfrac{1}{2}\sinh^{-1} t) \right]_0^{2\pi}$$
$$= \pi\sqrt{1 + 4\pi^2} + \tfrac{1}{2}\sinh^{-1} 2\pi \,.$$

5.5. Writing $\gamma(t) = e^{t+it}$, we have $\gamma'(t) = (1 + i)e^{t+it}$ and so $|\gamma'(t)| = \sqrt{2}e^t$. Hence $\Lambda(\gamma^*) = \int_a^b \sqrt{2}e^t \, dt = \sqrt{2}(e^b - e^a)$, which tends to $\sqrt{2}e^b$ as $a \to -\infty$.

5.6. We use Theorem 5.19, with $F(\zeta) = -1/(n\zeta^n)$. Thus $\int_\gamma (d\zeta/\zeta^{n+1}) = \left[-1/(n\zeta^n) \right]_{z-a}^{z-a-h} = (1/n)[1/(z - a - h)^n - 1/(z - a)^n]$.

5.7. a) $\int_\gamma f(z) \, dz = \int_0^1 t^2(2t + i) \, dt = \left[(t^4/2) + i(t^3/3) \right]_0^1 = (1/6)(3 + 2i)$.
    b) Here we can use Theorem 5.19, noting that $z = 0$ when $t = 0$ and $z = -1$ when $t = \pi$: so $\int_\gamma f(z) \, dz = \left[z^3/3 \right]_1^{-1} = -2/3$.
    c) This is not a simple curve, since, for example, $\gamma(2\pi) = \gamma(4\pi)$. So, from the definition, $\int_\gamma f(z) \, dz = \int_0^{6\pi} e^{-it} \cdot ie^{it} \, dt = 6\pi i$.
    d) Since the curve is piecewise smooth, we may use Theorem 5.19, with $F(z) = \sin z$. Thus $\int_\gamma \cos z \, dz = \sin(\pi + i\pi) - \sin(-\pi - i\pi) = 2\sin(\pi + i\pi) = 2(\sin\pi\cosh\pi + i\cos\pi\sinh\pi) = -2i\sinh\pi$.

5.8. $|z^4| = \left(|z|^2 \right)^2 = [(1 - t)^2 + t^2]^2 = (2t^2 - 2t + 1)^2 = 4[(t - \tfrac{1}{2})^2 + \tfrac{1}{4}]^2 \geq \tfrac{1}{4}$. Since γ^* has length $\sqrt{2}$, it follows by Theorem 5.24 that $|I| \leq 4\sqrt{2}$. We evaluate I using Theorem 5.19, with $F(z) = -1/(3z^3)$. Thus $I = \left[-1/3z^3 \right]_i^1 = -\tfrac{1}{3} + \tfrac{i}{3}$, and so $|I| = \sqrt{2}/3$.

5.9. The result follows from the fact that, on the curve γ^*, $|z^3 - 4z + 1| \leq |z|^3 + |z| + 1 = R^3 + 4R + 1$, $|z^2 + 5| \geq |z|^2 - 5 = R^2 - 5$, and $|z^3 - 3| \geq$

$|z|^3 - 3 = R^3 - 3$. Thus $|f(z)| \leq (R^3 + 4R + 1)/[(R^2 - 5)(R^3 - 3)$. Since $\Lambda(\gamma^*) = \pi R$, we deduce from Theorem 5.24 that $\left|\int_\gamma f(z)\,dz\right| \leq [\pi R(R^3 + 4R + 1)]/[(R^2 - 5)(R^3 - 3)]$.

5.10.

$$
\begin{aligned}
|\sin(u + iv)|^2 &= |\sin u \cosh v + i \cos u \sinh v|^2 \\
&= \sin^2 u \cosh^2 v + \cos^2 u \sinh^2 v \\
&\leq \cosh^2 v + \sinh^2 v = \frac{1}{4}(e^v + e^{-v})^2 + \frac{1}{4}(e^v - e^{-v})^2 \\
&= \frac{1}{2}(e^{2v} + e^{-2v}) = \cosh 2v.
\end{aligned}
$$

5.11. Although the conditions for Theorem 5.19 are satisfied, we cannot write down a function F with the property that $F'(z) = \sin(z^2)$. But this matters not at all, since all we are looking for is an estimate. First, it is clear that the length of γ^* is $6a$. Next, from the previous example,

$$
|\sin(z^2)| = |\sin[(x^2 - y^2) + 2ixy]| \leq \cosh(4xy).
$$

The largest value of $\cosh(4xy)$ is obtained when $|4xy|$ is as large as possible, and this occurs at the corners $(\pm a, \pm a)$. Hence, by Theorem 5.24, $\left|\int_\gamma \sin(z^2)\,dz\right| \leq 6a \cosh(4a^2)$.

Chapter 6

6.1. By Theorem 6.7, the integral round γ is the same as the integral round the unit circle, and we know this integral from Theorem 5.13. If $\gamma(t) = a\cos t + ib\sin t$ $(0 \leq t \leq 2\pi)$, then

$$
\begin{aligned}
2\pi i = \int_\gamma \frac{1}{z}\,dz &= \int_0^{2\pi} \frac{-a\sin t + ib\cos t}{a\cos t + ib\sin t}\,dt \\
&= \int_0^{2\pi} \frac{(-a\sin t + ib\cos t)(a\cos t - ib\sin t)}{a^2\cos^2 t + b^2\sin^2 t}\,dt \\
&= \int_0^{2\pi} \frac{(b^2 - a^2)\sin t \cos t + iab(\cos^2 t + \sin^2 t)}{a^2\cos^2 t + b^2\sin^2 t}\,dt.
\end{aligned}
$$

Equating imaginary parts gives

$$
\int_0^{2\pi} \frac{dt}{a^2\cos^2 t + b^2\sin^2 t} = \frac{2\pi}{ab}.
$$

6.2. By Theorem 6.7, the integral round the contour shown in Exercise 5.11 is equal to the integral along the straight line from $(-a, a)$ to (a, a). The length of this line is $2a$, and so

$$\left| \int_\gamma \sin(z^2)\, dz \right| \leq 2a \cosh(4a^2).$$

6.3.

$$0 = \int_{\kappa(0,r)} e^z\, dz = \int_0^{2\pi} e^{r(\cos\theta + i\sin\theta)} i r e^{i\theta}\, d\theta$$

$$= \int_0^{2\pi} i r e^{r\cos\theta} e^{i(\theta + r\sin\theta)}\, d\theta$$

$$= ir \int_0^{2\pi} e^{r\cos\theta} [\cos(\theta + r\sin\theta) + i\sin(\theta + r\sin\theta)]\, d\theta.$$

Dividing by ir and taking real parts gives the required result.

Chapter 7

7.1. a) From (7.5), $\int_{\kappa(0,1)} \left(e^{kz}/z^{n+1} \right) dz = (2\pi i/n!) f^{(n)}(0)$, where $f(z) = e^{kz}$, and this is equal to $2\pi i k^n/n!$.

   b) Since $1/(z^2 - 2z + 2) = (i/2)\left[\left(1/[z - (1-i)]\right) - \left(1/[z - (1+i)]\right)\right]$, we can split the integral in two and obtain the value $-\pi[(1-i)^3 - (1+i)^3] = 4\pi i$.

   c) The integral is equal to $-(1/2) \int_{\kappa(0,2)} e^z/[z - (\pi i/2)]\, dz$, and this is equal to $-\pi i e^{\pi i/2} = \pi$.

7.2. $\int_{\kappa(0,2)} [z^m/(1-z)^n]\, dz = (-1)^n \int_{\kappa(0,2)} [z^m/(z-1)^n]\, dz = (-1)^n [2\pi i/(n-1)!]m(m-1)\ldots(m-n+2) = 2\pi i \binom{m}{n-1}$. This is equal to zero if $m < n - 2$.

7.3.

$$f'(a) = \frac{1}{2\pi i} \int_{\kappa(a,r)} \frac{f(z)}{(z-a)^2}\, dz = \frac{1}{2\pi i} \int_0^{2\pi} \frac{f(a + re^{i\theta})}{r^2 e^{2i\theta}} i r e^{i\theta}\, d\theta$$

$$= \frac{1}{2r\pi} \int_0^{2\pi} f(a + re^{i\theta}) e^{-i\theta}\, d\theta.$$

Suppose that $\mathrm{Re}[f(a + re^{i\theta})] = F$ and $\mathrm{Im}[f(a + re^{i\theta})] = G$. Thus $f'(a) = J + iK$, where $J = (1/2r\pi) \int_0^{2\pi} [F\cos\theta + G\sin\theta]\, d\theta = F_c + G_s$

(by splitting the integral into two and using an obvious notation), and $K = (1/2r\pi) \int_0^{2\pi} [-F\sin\theta + G\cos\theta] = -F_s + G_c$. That is,

$$f'(a) = F_c + G_s + i(-F_s + G_c). \qquad (13.2)$$

Now, the integral $L = (1/2r\pi) \int_0^{2\pi} f(a + re^{i\theta})e^{i\theta} d\theta$ is equal to $(1/2r^2\pi i) \int_{\kappa(a,r)} f(z) dz$, and this equals 0, since f is holomorphic. That is,

$$\frac{1}{2r\pi} \int_0^{2\pi} [F\cos\theta - G\sin\theta] d\theta = \frac{1}{i} 2r\pi \int_0^{2\pi} [F\sin\theta + G\cos\theta] = 0.$$

In the notation of (13.2), we have

$$F_c - G_s = 0, \quad F_s + G_c = 0, \qquad (13.3)$$

and from (13.2) and (13.3) we deduce that

$$f'(a) = 2(F_c - iG_c) = \frac{1}{\pi r} \int_0^{2\pi} \text{Re}[f(a + re^{i\theta})]e^{-i\theta} d\theta.$$

7.4. a) Write $g(z) = (R^2 - a\bar{a})f(z)/(R^2 - z\bar{a})$. Since $|z\bar{a}| < R^2$, the denominator is non-zero inside and on the contour. Hence g is holomorphic, and $(1/2\pi i) \int_{\kappa(0,R)} [g(z)/(z - a)] = g(a) = f(a)$.

   b) Hence, writing $z = Re^{i\phi}$ and $a = re^{i\theta}$, we see that

$$f(a) = \frac{1}{2\pi i} \int_0^{2\pi} \frac{(R^2 - r^2)f(Re^{i\phi})iRe^{i\phi} d\phi}{(R^2 - Rre^{i(\phi-\theta)})(Re^{i\phi} - re^{i\theta})}$$

$$= \frac{1}{2\pi} \int_0^{2\pi} \frac{(R^2 - r^2)f(Re^{i\phi}) d\phi}{(R - re^{i(\phi-\theta)})(R - re^{i(\theta-\phi)})}$$

$$= \frac{1}{2\pi} \int_0^{2\pi} \frac{(R^2 - r^2)}{R^2 - 2Rr\cos(\theta - \phi) + r^2} f(Re^{i\phi}) d\phi.$$

7.5. $1/([z - (\pi/6)]^2[z + (\pi/6)]) = A/[z - (\pi/6)] + B/([z - (\pi/6)]^2) + C/[z + (\pi/6)]$, where $A = -9/\pi^2$, $B = 3/\pi$ and $C = 9/\pi^2$. Hence $I = (-9/\pi^2) \int_{\kappa(0,4)} \sin^2 z \, dz/[z - (\pi/6)] + (3/\pi) \int_{\kappa(0,4)} \sin^2 z \, dz/[z - (\pi/6)]^2 + (9/\pi^2) \int_{\kappa(0,4)} \sin^2 z \, dz/[z + (\pi/6)] = (-9/\pi^2)2\pi i \sin^2(\pi/6) + (3/\pi)2\pi i[2\sin(\pi/6)\cos(\pi/6)] + (9/\pi^2)2\pi i \sin^2(-\pi/6) = 3i\sqrt{3}$.

7.6. The continuous function f is bounded on γ^*: say $|f(z)| \leq M$. For all z in $I(\gamma)$ there exists $N(z, \delta)$ wholly contained in $\mathbb{C} \setminus \gamma^*$. Thus $|z - w| > \delta$ for every point w on γ^*. If h is such that $|h| < \delta/2$, then $|z - w - h| > \delta/2$ for every point w on γ^*.

Note now that

$$\frac{1}{h}[g(z+h) - g(z)] = \frac{1}{h}\int_\gamma f(w)\left(\frac{1}{w-z-h} - \frac{1}{w-z}\right) dw$$

$$= \int_\gamma \frac{f(w)\,dw}{(w-z)(w-z-h)}.$$

Hence

$$\left|\frac{1}{h}[g(z+h) - g(z)] - \int_\gamma \frac{f(w)}{(w-z)^2}\,dw\right|$$

$$= \left|\int_\gamma f(w)\left(\frac{1}{(w-z)(w-z-h)} - \frac{1}{(w-z)^2}\right) dw\right|$$

$$= |h|\left|\int_\gamma \frac{f(w)}{(w-z)^2(w-z-h)}\,dw\right| \le |h|\frac{2M\Lambda(\gamma^*)}{\delta^3},$$

which tends to 0 as $h \to 0$. Hence the derivative of g exists and equals $\int_\gamma [f(w)/(w-z)^2]\,dw$.

7.7.   By the Fundamental Theorem of Algebra, $p(x)$ factorises as

$$a_n(x - \alpha_1)(x - \alpha_2)\ldots(x - \alpha_n).$$

Suppose that the roots are ordered so that $\alpha_1, \ldots, \alpha_k$ $(k \ge 0)$ are real, and $\alpha_{k+1}, \ldots, \alpha_n \in \mathbb{C} \setminus \mathbb{R}$. As observed in Exercise 2.18, the remaining factors occur in conjugate pairs $x - \mu$, $x - \bar\mu$, and so $l = n - k$ is even. The two factors combine to give a real quadratic factor $x^2 - 2\operatorname{Re}\mu + |\mu|^2$. If n is odd, then $k = n - l$ must also be odd, and so is at least 1.

7.8. $x^6 + 1 = (x - e^{\pi i/6})(x - e^{-\pi i/6})(x - e^{3\pi i/6})(x - e^{-3\pi i/6})(x - e^{5\pi i/6})(x - e^{-5\pi i/6}) = (x^2 - 2x\cos(\pi/6) + 1)(x^2 + 1)(x^2 - 2x\cos(5\pi/6) + 1) = (x^2 + 1)(x^2 - x\sqrt{3} + 1)(x^2 + x\sqrt{3} + 1)$. Here $k = 0$.
$x^4 - 3x^3 + 4x^2 - 6x + 4 = (x - 1)(x - 2)(x^2 + 2)$.
$x^4 + 3x^3 - 3x^2 - 7x + 6 = (x - 1)^2(x + 2)(x + 3)$. Here $l = 0$.

7.9.   If f is even, then $0 = f(z) - f(-z) = (a_0 + a_1 z + a_2 z^2 + \cdots) - (a_0 - a_1 z + a_2 z^2 - \cdots) = 2(a_1 z + a_3 z^3 + \cdots)$. This is the unique Taylor series for the zero function, and so coincides with the obvious Taylor series $0 + 0z + 0z^2 + \cdots$. Hence $a_{2n+1} = 0$ for all $n \ge 0$. The odd function is dealt with in the same way.

7.10.

$$e^z = e^c e^{z-c} = e^c\left[1 + (z-c) + \frac{(z-c)^2}{2!} + \frac{(z-c)^3}{3!} + \cdots\right].$$

$$\cos z = \cos[(z-c)+c] = \cos(z-c)\cos c - \sin(z-c)\sin c$$
$$= \cos c \sum_{n=0}^{\infty} \frac{(-1)^n}{(2n)!} (z-c)^{2n} - \sin c \sum_{n=0}^{\infty} \frac{(-1)^n}{(2n+1)!} (z-c)^{2n+1}.$$

7.11. a) From (7.7), $a_n = (1/2\pi i) \int_{\kappa(0,r)} [f(z)/z^{n+1}]\, dz$; hence

$$|a_n| \le (1/2\pi) . 2\pi r . M(r)/r^{n+1} = M(r)/r^n . \qquad (13.4)$$

b) Since $|f(z)| \le M$ for some M, $|a_n| \le M/r^n$ for all r. Letting $r \to \infty$, we see that $a_n = 0$ for all $n \ge 1$. Thus f is constant.

c) From (13.4) we have $|a_N| \le Kr^{N-n}$ for all r. Letting $r \to \infty$, we see that $a_n = 0$ for all $n > N$. Thus $f(z)$ is a polynomial of degree at most N.

7.12. By Leibniz's formula,

$$c_n = \frac{1}{n!} h^{(n)}(0) = \sum_{r=0}^{n} \frac{f^{(n-r)}(0)}{(n-r)!} \cdot \frac{g^{(r)}(0)}{r!} = \sum_{r=0}^{n} a_{n-r} b_r .$$

7.13. The series for sin and cos give the identity

$$z - \frac{z^3}{3!} + \frac{z^5}{5!} - \cdots = (a_1 z + a_3 z^3 + \cdots)\left(1 - \frac{z^2}{2!} + \frac{z^4}{4!} - \cdots\right).$$

From the previous example, it follows, by equating coefficients of z^{2n+1}, that

$$\frac{(-1)^n}{(2n+1)!} = a_{2n+1} - \frac{a_{2n-1}}{2!} + \frac{a_{2n-3}}{4!} - \cdots + (-1)^n \frac{a_1}{(2n)!} .$$

Putting $n = 0$ gives $a_1 = 1$. Putting $n = 1$ gives $a_3 - (a_1/2) = -1/6$, and so $a_3 = 1/3$. Putting $n = 2$ gives $a_5 - (a_3/2) + (a_1/24) = 1/120$, and a routine calculation gives $a_5 = 2/15$.

7.14. From the definitions, $\tanh z = -i\tan(iz) = -i(a_1(iz) + a_3(iz)^3 + a_5(iz)^5 + \cdots) = a_1 z - a_3 z^3 + a_5 z^5 - \cdots$. In general, $b_{2n+1} = (-1)^n a_{2n+1}$.

Chapter 8

8.1. $1/\sin z = z^{-1}[1 - \frac{1}{6}z^2 + \frac{1}{120}z^4 + o(z^5)]^{-1} = z^{-1}\left[1 + \left(\frac{1}{6}z^2 - \frac{1}{120}z^4 + o(z^5)\right) + \left(\frac{1}{6}z^2 - \frac{1}{120}z^4 + o(z^5)\right)^2 + o(z^5)\right] = z^{-1}[1 + \frac{1}{6}z^2 + (\frac{1}{36} - \frac{1}{120})z^4 + o(z^5)] = z^{-1} + \frac{1}{6}z + \frac{7}{360}z^3 + o(z^4).$

8.2. $(1 - \cos z)^{-1} = [\frac{1}{2}z^2 - \frac{1}{24}z^4 + \frac{1}{720}z^6 o(z^7)]^{-1} = 2z^{-2}[1 - \frac{1}{12}z^2 + \frac{1}{360}z^4 + o(z^5)]^{-1} = 2z^{-2}[1 + (\frac{1}{12}z^2 - \frac{1}{360}z^4 + o(z^5)) + (\frac{1}{12}z^2 - \frac{1}{360}z^4 + o(z^5))^2 + o(z^5)] = 2z^{-2}[1 - \frac{1}{12}z^2 + (\frac{1}{144} - \frac{1}{360})z^4 + o(z^5)] = 2z^{-2} + \frac{1}{6} + \frac{1}{120}z^2 + o(z^3).$

8.3. $e^{1/z}e^{2z} = [1 + (1/z) + (1/2!)(1/z)^2 + (1/3!)(1/z)^3 + \cdots][1 + 2z + (1/2!)(2z)^2 + (1/3!)(2z)^3 + \cdots]$. The coefficient of z^{-1} is $1 + (1/2!)2 + (1/3!)(2^2/2!) + (1/4!)(2^3/3!) + \cdots = \sum_{n=0}^{\infty}(2^n/[n!(n+1)!])$.

8.4.   a) From Exercise 8.1, $1/(z^4 \sin z) = z^{-5} + \frac{1}{6}z^{-3} + \frac{7}{360}z^{-1} + o(1)$, and so $\operatorname{res}(f, 0) = \frac{7}{360}$.

   b) From Exercise 8.2, $1/[z^3(1 - \cos z)] = 2z^{-5} + \frac{1}{6}z^{-3} + \frac{1}{120}z^{-1} + o(1)$, and so $\operatorname{res}(f, 0) = \frac{1}{120}$.

8.5.   a) Let $\operatorname{ord}(f, c) = m$, $\operatorname{ord}(g, c) = n$, so that $f(z) = (z - c)^m f^*(z)$, $g(z) = (z - c)^n g^*(z)$, where f^* and g^* are differentiable and non-zero at c. Hence $(f \cdot g) = (z - c)^{m+n}(f^* \cdot g^*)(z)$, where $f^* \cdot g^*$ is differentiable and non-zero at c. Thus $\operatorname{ord}(f \cdot g, c) = m + n = \operatorname{ord}(f, c) + \operatorname{ord}(g, c)$.

   b) With the same notation, $(1/f)(z) = (z - c)^{-m}(1/f^*)(c)$. Since $1/f^*$ is differentiable and non-zero at c, it follows that $\operatorname{ord}(1/f\, c) = -m = -\operatorname{ord}(f, c)$.

   c) Suppose that $m < n$. Then $(f + g)(z) = (z - c)^m h(z)$, where $h(z) = f^*(z) + (z - c)^{n-m} g^*(z)$. Since h is differentiable at c, and since $h(c) = f^*(c) \neq 0$, it follows that $\operatorname{ord}(f + g) = m = \operatorname{ord}f$.

8.6.   a) Since $\operatorname{ord}(1/\sin z, 0) = \operatorname{ord}(1/z, 0) = -1$, it follows from (a) above that $\operatorname{ord}(1/z \sin^2 z, 0) = -3$. That is, the function has a triple pole at 0.

   b) From $\operatorname{ord}(\cot z, 0) = -1$, $\operatorname{ord}(\cos z, 0) = 0$, $\operatorname{ord}(\sin 2z, 0) = 1$, we deduce that $\operatorname{ord}([(\cot z + \cos z)/\sin 2z], 0) = -2$.

   c) Clearly $\operatorname{ord}(z^2(z - 1), 0) = 2$. Since $\log(1 + z) = z - \frac{1}{2}z^2 + \cdots$ and $1 - \cos z = \frac{1}{2}z^2 - \frac{1}{24}z^4 + \cdots$, we have that $\operatorname{ord}((1 - \cos z)\log(1 + z), 0) = 3$. Hence $\operatorname{ord}(z^2(z - 1)/[(1 - \cos z)\log(1 + z)], 0) = 2 - 3 = -1$.

8.7.   Suppose that the Laurent series of f at zero is $f(z) = \cdots + a_{-3}z^{-3} + a_{-2}z^{-2} + a_{-1}z^{-1} + a_0 + a_1 z + \cdots$. Then $f(-z) = \cdots - a_{-3}z^{-3} + a_{-2}z^{-2} - a_{-1}z^{-1} + a_0 - a_1 z + \cdots$, and so from $f(z) = f(-z)$ we deduce that $0 = \cdots + a_{-3}z^{-3} + a_{-1}z^{-1} + a_1 z + \cdots$. By the uniqueness theorem this must coincide with the obvious Laurent series $\sum_{n=-\infty}^{\infty} 0z^n$ for the zero function. Hence $a_n = 0$ for all odd n. In particular, the residue a_{-1} is zero.

8.8. $z - \sin z = (z^3/6) - (z^5/120) + \cdots = (1/6)z^3[1 - (z^2/20) + O(z^4)]$. Hence $1/(z - \sin z) = (6/z^3)[1 - (z^2/20) + O(z^4)]^{-1} = (6/z^3)[1 + (z^2/20) + O(z^4)]$. Hence $\operatorname{res}((z - \sin z)^{-1}, 0) = 3/10$.

8.9. The function has a double pole at -1, and $f(z) = (z+1)^{-2}g(z)$, where
      $g(z) = 1/(z^2 - z + 1)^2$. Then $g'(z) = -2(2z - 1)/(z^2 - z + 1)^3$, and so
      $\mathrm{res}(f, -1) = g'(-1) = 6/27 = 2/9$.
      Alternatively, writing $z^3 + 1 = [(z+1) - 1]^3 + 1 = (z+1)^3 - 3(z+1)^2 + 3(z+1)$, we see that $f(z) = (1/9)(z+1)^{-2}[1 - (z+1) + (1/3)(z+1)^2]^{-2} = (1/9)(z+1)^{-2}(1 + 2(z+1) + O((z+1)^2))$ The residue is the coefficient of
      $(z+1)^{-1}$, namely $2/9$.

8.10. $(1 + z^6)z^{-3}(1 - 2z)^{-1}(1/2)(1 - \frac{1}{2}z)^{-1} = \frac{1}{2}z^{-3}(1 + z^6)[1 + 2z + 4z^2 + O(z^3)][1 + \frac{1}{2}z + \frac{1}{4}z^2 + O(z^3)] = \frac{1}{2}z^{-3}[1 + \frac{5}{2}z + \frac{21}{4}z^2 + O(z^3)]$. Hence the
      residue is $21/8$.

8.11. From Example 7.16 we know that $\cot \pi z = (1/\pi z) - (\pi z)/3 + O(z^3)$.
      Hence $\cot \pi z/z^2 = (1/\pi z^3) - (\pi/3z) + O(z)$, and so the residue is $-\pi/3$.

8.12. By the periodic properties of the circular functions, $\cot \pi(z - n) = \cot \pi z$
      and $\operatorname{cosec} \pi(z - n) = (-1)^n \operatorname{cosec} \pi z$. Hence $(z - n)\cot \pi z = \cos \pi(z - n)[(z-n)/\sin \pi(z-n)] \to 1/\pi$ as $z \to n$, and $(z-n)\operatorname{cosec} \pi z = (-1)^n[(z-n)/\sin \pi(z-n)] \to (-1)^n/\pi$ as $z \to n$.
      If f has no zeros on the x-axis, $\pi f(z) \cot \pi z$ has a simple pole at each
      integer n, and the residue is $\pi f(n) \lim_{z \to n}(z - n)\cot \pi z = f(n)$. The
      other result follows in the same way.

8.13. a) There is a triple pole at $z = -1/2$. We find the Laurent series at
         $1/2$, noting first that $\sin \pi z = \sin[\pi(z + \frac{1}{2}) - \frac{1}{2}\pi] = -\cos[\pi(z + \frac{1}{2})] = -1 + \frac{1}{2}\pi^2(z + \frac{1}{2})^2 + O((z + \frac{1}{2})^4)$. Hence $\sin \pi z/(2z + 1)^3 = \frac{1}{8}(z + \frac{1}{2})^{-3}[-1 + \frac{1}{2}\pi^2(z + \frac{1}{2})^2 + O((z + \frac{1}{2})^4)]$, and so the residue at $-1/2$ is
         $\pi^2/16$. By the Residue Theorem, the value of the integral is $\pi^3 i/8$.
      b) There is a triple pole at 0, and $1/z^2 \tan z = (1/z^2)[1 + \frac{1}{3}z + O(z^3)]$.
         The residue at 0 is $1/3$, and so the integral has value $2\pi i/3$.

Chapter 9

9.1. Inside the contour $\sigma(0, R)$, $z^2/(1+z^4)$ has simple poles at $e^{i\pi/4}$ and $e^{3\pi i/4}$,
      with residues $1/4e^{i\pi/4} = (1 - i)/4\sqrt{2}$ and $1/4e^{3i\pi/4} = (-1 - i)/4\sqrt{2}$,
      respectively. Since the conditions of Theorem 9.1 are satisfied,

      $$\int_{-\infty}^{\infty} [x^2/(1 + x^4)]\, dx = [(2\pi i)/4\sqrt{2}](-2i) = \pi/\sqrt{2}\,.$$

      Hence
      $$\int_{0}^{\infty} [x^2/(1 + x^4)]\, dx = \pi/(2\sqrt{2})\,.$$

9.2. Inside $\sigma(0, R)$ there are simple poles at i and $2i$, with residues $-1/(6i)$, $(-4)/[(-3)(4i)] = 1/3i$, respectively. Hence

$$\int_{-\infty}^{\infty} [x^2/[(x^2 + 1)(x^2 + 4)]]\, dx = 2\pi i[-1/(6i) + 1/(3i)] = \pi/3.$$

9.3. Inside $\sigma(0, R)$ there are simple poles at $e^{i\pi/6}$, i and $e^{5i\pi/6}$, with residues (calculated with the aid of Theorem 8.15) $(-\sqrt{3}-i)/12$, $-i/6$, $(\sqrt{3}-i)/12$, respectively. Hence $\int_{-\infty}^{\infty} [1/(x^6 + 1)]\, dx = (2\pi i)(-i/3) = 2\pi/3$. Since the function is even, it follows that $\int_0^{\infty} [1/(x^6 + 1)]\, dx = \pi/3$.

9.4. Inside $\sigma(0, R)$ there is a triple pole at $z = i$, with residue

$$\frac{1}{2}\left[\frac{d^2}{dz^2}\left(\frac{1}{(z + i)^3}\right)\right]_{z=i} = \frac{6}{(2i)^5} = -\frac{3i}{16}.$$

Hence $\int_{-\infty}^{\infty} [1/(x^2 + 1)^3]\, dx = 2\pi i[-3i/16] = 3\pi/8$.

9.5. There is a pole of order n at $z = i$. The residue is

$$\frac{1}{(n - 1)!}\left[\frac{d^{n-1}}{dz^{n-1}}\left(\frac{1}{(z + i)^n}\right)\right]_{z=i}$$

$$= \frac{1}{(n - 1)!}\left[\frac{(-n)(-n - 1)\ldots(-n - n + 2)}{(z + i)^{2n-1}}\right]_{z=i}$$

$$= \frac{(-1)^{n-1}n(n + 1)\ldots(2n - 2)(-1)^n i}{2^{2n-1}(n - 1)!} = -\frac{i}{2^{2n-1}}\binom{2n - 2}{n - 1},$$

and so

$$\int_{-\infty}^{\infty} \frac{1}{(x^2 + 1)^n}\, dx = \frac{\pi}{2^{2n-2}}\binom{2n - 2}{n - 1}.$$

9.6. Since $z^2 + z + 1 = (z - e^{2\pi/3})(z - e^{-2\pi i/3})$, the function has a double pole inside $\sigma(0, R)$ at $e^{2\pi i/3}$, with residue

$$\left[\frac{d}{dz}\left(\frac{1}{(z - e^{-2\pi i/3})^2}\right)\right]_{z=e^{2\pi i/3}} = -\frac{2}{(2i\sin(2\pi/3))^3} = -\frac{2i}{3\sqrt{3}}.$$

So $\int_{-\infty}^{\infty} [1/(x^2 + x + 1)^2]\, dx = 4\pi/3\sqrt{3}$.

9.7. Consider the function $f(z) = e^{iz}/(z^2 + 1)$, and use the contour $\sigma(0, R)$, with $R > 1$. The function has a simple pole at i, with residue $e^{-1}/2i$. The modulus of the contribution of the circular arc is

$$\left|\int_0^{\pi} \frac{e^{-R\sin\theta + iR\cos\theta}iRe^{i\theta}}{R^2 e^{2i\theta} + 1}\, d\theta\right| \le \frac{\pi R}{R^2 - 1}$$

since $\sin\theta$ is non-negative for θ in $[0, \pi]$. This tends to 0 as $R \to \infty$. Letting $R \to \infty$ we see that $\int_{-\infty}^{\infty} [e^{ix}/(x^2 + 1)]\, dx = 2\pi i e^{-1}/(2i) =$

π/e. Taking real parts gives $\int_{-\infty}^{\infty} [\cos x/(x^2 + 1)]\,dx = \pi/e$, and so $\int_0^{\infty} [\cos x/(x^2 + 1)]\,dx = \pi/(2e)$, since the integrand is an even function.

9.8.   Integrate $e^{2iz}/(z^2 + 1)$ round the contour $\sigma(0, R)$. There is one simple pole i, with residue $e^{-2}/(2i)$. On the semicircular arc,

$$\left| \int_0^{\pi} \frac{e^{-2R\sin\theta + 2iR\cos\theta} iRe^{i\theta}\,d\theta}{R^2 e^{2i\theta} + 1} \right| \leq \frac{\pi R}{R^2 - 1},$$

which tends to 0 as $R \to \infty$. Hence $\int_{-\infty}^{\infty} [e^{2ix}\,dx/(x^2 + 1)] = 2\pi i/(2ie^2) = \pi/e^2$. Taking real parts, and using the even function property gives $\int_0^{\infty} [\cos 2x/(x^2 + 1)]\,dx = \pi/(2e^2)$. Finally, $\int_0^{\infty} [\sin^2 x/(x^2 + 1]\,dx = \frac{1}{2}\int_0^{\infty} [1/(x^2 + 1)]\,dx - \frac{1}{2}\int_0^{\infty} [\cos 2x/(x^2 + 1)]\,dx = (\pi/4)(1 - e^{-2})$.

9.9.   The function $f(z) = e^{iz}/[(z^2 + c^2)(z^2 + d^2)]$ has poles at ci, di in the upper half-plane, with residues $e^{-c}/[-2ic(c^2 - d^2)]$, $e^{-d}/[2id(c^2 - d^2)]$. In the upper half-plane, with $z = x + iy$, $|zf(z)| \leq e^{-y}/[(|z|^2 - c^2)(|z|^2 - d^2)]$, and so tends to 0 as $|z| \to \infty$. Hence, by Theorem 9.1, $\int_{-\infty}^{\infty} f(z)\,dz = [\pi/(c^2 - d^2)][(e^{-d}/d) - (e^{-c}/c)]$.

9.10.   The function $f(z) = e^{iz}/(z^2 + c^2)^2$ has a double pole at ci, with residue $g'(ic)$, where $g(z) = e^{iz}/(z + ic)^2$. A routine calculation gives $g'(ic) = e^{-c}(c+1)/(4ic^2)$. As in the previous exercise, the conditions of Theorem 9.1 are satisfied. Hence $\int_{-\infty}^{\infty} f(z)\,dz = \pi(c + 1)e^{-c}/(2c^3)$.

9.11.   Integrate $e^{isz}/(k^2 + z^2)$ round the semicircular contour $\sigma(0, R)$. For the contribution of the curved part,

$$\left| \int_0^{\pi} \frac{e^{is(R\cos\theta + iR\sin\theta)} iRe^{i\theta}\,d\theta}{k^2 + R^2 e^{2i\theta}} \right| \leq \frac{\pi R}{R^2 - k^2},$$

since $\sin\theta \geq 0$ in $[0, \pi]$, and this tends to 0 as $R \to \infty$. The integrand has a pole ki within $\sigma(0, R)$, with residue $e^{-ks}/2ki$. Hence

$$\int_{-\infty}^{\infty} [e^{isx}/(k^2 + x^2)]\,dx = (\pi/k)e^{-ks}.$$

Then equate real parts.

9.12.   Substituting $z = e^{i\theta}$ gives $I = (1/i)\int_{\kappa(0,1)} [1/(-az^2 + (1 + a^2)z - a)]\,dz = (1/i)\int_{\kappa(0,1)} \left(1/[(z - a)(1 - az)]\right)\,dz$. If $|a| < 1$ there is a pole inside $\kappa(0, 1)$ at a, with residue $1/(1 - a^2)$, and so $I = 2\pi/(1 - a^2)$

If $|a| > 1$ the pole inside $\kappa(0, 1)$ is at $1/a$, and the residue is $\lim_{z \to 1/a}[z - (1/a)]/[(z - a)(az - 1)] = 1/(a^2 - 1)$. Hence $I = 2\pi/(a^2 - 1)$.)

9.13. On substituting $z = e^{i\theta}$ we transform the given integral to

$$I = \frac{1}{4} \int_{\kappa(0,1)} \frac{(z^2 + z^{-2})^2}{(1 - pz)(1 - pz^{-1})} \frac{1}{iz} \, dz = \frac{1}{4i} \int_{\kappa(0,1)} \frac{(z^4 + 1)^2}{z^4(1 - pz)(z - p)} \, dz.$$

Within the contour the integrand has a simple pole at p and a pole of order 4 at 0. The residue at p is $(p^4 + 1)^2/[p^4(1 - p^2)]$. Near 0 we have the Laurent series

$$\frac{1}{z^4}(1 + 2z^4 + z^8)(1 + pz + p^2z^2 + \cdots)(-\frac{1}{p})(1 + \frac{z}{p} + \frac{z^2}{p^2} + \cdots),$$

and the coefficient of z^{-1} is

$$\left(-\frac{1}{p}\right)\left(p^3 + p + \frac{1}{p} + \frac{1}{p^3}\right) = -\frac{1}{p^4}(p^6 + p^4 + p^2 + 1) = -\frac{1 - p^8}{p^4(1 - p^2)}.$$

So

$$I = \frac{\pi}{2}\left(\frac{p^8 + 2p^4 + 1 - 1 + p^8}{p^4(1 - p^2)}\right) = \frac{\pi(1 + p^4)}{1 - p^2}.$$

9.14. The substitution $z = e^{i\theta}$ transforms the integral to

$$-\frac{1}{2i} \int_{\kappa(0,1)} \frac{(z^2 - 1)^2 \, dz}{z^2(bz^2 + 2az + b)} = -\frac{1}{2bi} \int_{\kappa(0,1)} \frac{(z^2 - 1)^2 \, dz}{(z - \alpha)(z - \beta)},$$

where $\alpha = (-a + \sqrt{a^2 - b^2})/b$ and $\beta = (-a - \sqrt{a^2 - b^2})/b$ are the roots of the equation $bz^2 + 2az + b = 0$. Note that $\alpha\beta = 1$. Since $|\beta| = (a + \sqrt{a^2 - b^2})/b > a/b > 1$, it follows that $|\alpha| < 1$, and so the only relevant poles are a simple pole at α and a double pole at 0. The residue at α is $(\alpha^2 - 1)^2/[\alpha^2(\alpha - \beta)]$. Now, $(\alpha^2 - 1)/\alpha^2 = [\alpha - (1/\alpha)]^2 = (\alpha - \beta)^2$, and so it follows that the residue at α is $\alpha - \beta = 2\sqrt{a^2 - b^2}/b$. Near 0 the integrand has Laurent series

$$\frac{1}{z^2}(1 - 2z^2 + z^4)(1 + (2a/b)z + z^2)^{-1} = \frac{1}{z^2}(1 - 2z^2 + z^4)\left(1 - (2a/b)z + O(z^2)\right),$$

and it is clear that the coefficient of z^{-1} is $-2a/b$. Hence

$$I = \frac{\pi}{b}\left(\frac{2a}{b} - \frac{2\sqrt{a^2 - b^2}}{b}\right) = \frac{2\pi}{b^2}\left(a - \sqrt{a^2 - b^2}\right).$$

9.15. $I = \int_0^{2\pi} [e^{3i\theta}/(\cosh \alpha - \cos \theta)] \, d\theta = 2i \int_{\kappa(0,1)} [z^3/(z^2 - 2z \cosh \alpha + 1)] \, dz =$
$2i \int_{\kappa(0,1)} \left(z^3/[(z - e^\alpha)(z - e^{-\alpha})] \right) dz$. Since $\alpha > 0$ we have $0 < e^{-\alpha} < 1 < e^\alpha$. So the integrand has a simple pole within $\kappa(0,1)$ at $e^{-\alpha}$. The residue is $e^{-3\alpha}/(e^{-\alpha} - e^\alpha) = -(1/2)e^{-3\alpha}/\sinh \alpha$. Hence

$$I = 2\pi i \,.\, 2i \,.\, (-1/2)e^{-3\alpha}/\sinh \alpha = 2\pi e^{-3\alpha}/\sinh \alpha \,.$$

Taking real and imaginary parts gives this as the value of the required integral, while $\int_0^{2\pi} [\sin 3\theta/(\cosh \alpha - \cos \theta)] \, d\theta = 0$.
If $\cosh \alpha = 5/4$ then $\sinh \alpha = (\cosh^2 \alpha - 1)^{1/2} = 3/4$, and $e^{-\alpha} = \cosh \alpha - \sinh \alpha = 1/2$. Hence $\int_0^{2\pi} [\cos 3\theta/(5 - 4\cos \theta)] \, d\theta = (1/4)(2\pi)(1/8)(4/3) = \pi/12$.

9.16. Within $\kappa(0,1)$ there is a pole (of order $n+1$) at 0. The Laurent expansion is $(1/z^{n+1})(1 + z + \cdots + (z^n/n!) + \cdots)$, and the coefficient of z^{-1} is $1/n!$. Hence $\int_{\kappa(0,1)} [e^z/z^{n+1}] \, dz = 2\pi i/n!$. The integral is equal to

$$\int_0^{2\pi} [e^{\cos \theta + i \sin \theta} i e^{i\theta}/e^{(n+1)i\theta}] \, d\theta = i \int_0^{2\pi} e^{\cos \theta} e^{i(\sin \theta - n\theta)} \, d\theta$$

$$= i \int_0^{2\pi} e^{\cos \theta} \left(\cos(\sin \theta - n\theta) + i \sin(\sin \theta - n\theta) \right) d\theta \,.$$

Hence, equating imaginary parts, we have $\int_0^{2\pi} e^{\cos \theta} \cos(n\theta - \sin \theta) \, d\theta = 2\pi/n!$.

9.17. The integrand has a pole at ia, with residue e^{ima}. Hence the integral has the value $2\pi i(\cos ma + i \sin ma) = 2\pi(-\sin ma + i \cos ma)$. If we substitute $z = e^{i\theta}$, the integral becomes

$$\int_0^{2\pi} \frac{e^{m(\cos \theta + i \sin \theta)} i e^{i\theta} \, d\theta}{e^{i\theta} - ia} = i \int_0^{2\pi} \frac{e^{m(\cos \theta + i \sin \theta)} i e^{i\theta} (e^{-i\theta} + ia) \, d\theta}{1 - 2a \sin \theta + a^2}$$

$$= i \int_0^{2\pi} \frac{e^{m \cos \theta} [e^{im \sin \theta} + iae^{i(m \sin \theta + \theta)}] \, d\theta}{1 - 2a \sin \theta + a^2}$$

$$= \int_0^{2\pi} \frac{e^{m \cos \theta} [ie^{im \sin \theta} - ae^{i(m \sin \theta + \theta)}] \, d\theta}{1 - 2a \sin \theta + a^2} \,.$$

The real and imaginary parts are (respectively)

$$\int_0^{2\pi} \frac{e^{m \cos \theta} [-\sin(m \sin \theta) - a \cos(m \sin \theta + \theta)] \, d\theta}{1 - 2a \sin \theta + a^2} \,,$$

$$\int_0^{2\pi} \frac{e^{m \cos \theta} [\cos(m \sin \theta) - a \sin(m \sin \theta + \theta)] \, d\theta}{1 - 2a \sin \theta + a^2} \,,$$

and the required results follow immediately.

9.18. In the upper half plane, $f(z) = 1/(z^2 - 2z + 2)$ has just one simple pole, at $1 + i$. From $|f(z)| < 1/(|z|^2 - 2|z| - 2)$, we deduce that $|f(z)| \to 0$ as $|z| \to \infty$. Hence Jordan's Lemma applies. The residue of $g(z) = f(z)e^{i\pi z}$ at $1 + i$ is $e^{i\pi(1+i)}/[(1+i) - (1-i)] = -e^{-\pi}/2i$, and so, equating real and imaginary parts, we have

$$\int_{-\infty}^{\infty} \frac{\cos \pi x \, dx}{x^2 - 2x + 2} = -\pi e^{-\pi}, \qquad \int_{-\infty}^{\infty} \frac{\sin \pi x \, dx}{x^2 - 2x + 2} = 0.$$

9.19. The function $f(z) = z^3/(1+z^2)^2$ has a double pole in the upper half plane at i. From $|f(z)| \le |z|^3/(|z|^2 - 1)^2$ we deduce that $|f(z)| \to 0$ as $|z| \to \infty$. The residue of $g(z) = f(z)e^{iz}$ at i is $h'(i)$, where $h(z) = z^3 e^{iz}/(z+i)^2$. Since $h'(z) = iz^2(z^2 + 3)e^{iz}/(z+i)^3$, we have $\text{res}(g, i) = 1/4e$. By Jordan's Lemma, $\int_{-\infty}^{\infty} [x^3 e^{ix}/(1+x^2)^2] \, dx = \pi i/(2e)$, and equating imaginary parts gives the desired result.

9.20. We use the contour from the proof of Theorem 9.8, with a semicircular indent of radius r so as to avoid 0. The function $f(z) = (z^2 - a^2)/[z(z^2 + a^2)]$ satisfies the conditions of Jordan's Lemma. The residues of $g(z) = [(z^2 - a^2)e^{iz}]/[z(z^2 + a^2)]$ at the simple poles 0 and ia are -1 and e^{-a}, respectively. If we denote the integral of $g(z)$ round the indent (in the positive direction) by I_r, we conclude that $\int_{-\infty}^{-r} g(x) \, dx - I_r + \int_r^{\infty} g(x) \, dx = 2\pi i e^{-a}$. By Lemma 9.12, $\lim_{r \to 0} I_r = \pi i$. Hence $\int_{-\infty}^{\infty} g(x) \, dx = \pi i(2e^{-a} + 1)$, and the required result follows by taking imaginary parts.

9.21. Here the function $f(z) = 1/[z(1 - z^2)]$ has three poles, all on the real axis. We use the contour of Theorem 9.8, indented at -1, 0 and 1 with semi-circles I_{-1}, I_0 and I_1 of radius r, and, denoting $e^{i\pi z}/[z(1 - z^2)]$ by $g(z)$, deduce that $\int_{-\infty}^{-1-r} g(x) \, dx - I_{-1} + \int_{-1+r}^{-r} g(x) \, dx - I_0 + \int_r^{1-r} g(x) \, dx - I_{-1} + \int_{-\infty}^{-1-r} g(x) \, dx - I_1 + \int_{1+r}^{\infty} g(x) \, dx = 0$. (The negative signs arise because the three semicircular indents are traversed in the negative direction.) Thus $\int_{-\infty}^{\infty} g(x) \, dx = \lim_{r \to 0}(I_{-1} + I_0 + I_1)$. The residues of $g(z)$ at -1, 0 and 1 are $\frac{1}{2}$, 1 and $\frac{1}{2}$, respectively, and so it follows from Lemma 9.12 that $\int_{-\infty}^{\infty} g(x) \, dx = 2\pi i$. The result is now obtained by taking imaginary parts and by noting that the integrand is an even function.

9.22. The function $e^{az}/\cosh z$ has infinitely many simple poles at $z = (n + \frac{1}{2})\pi i$ $(n \in \mathbb{Z})$. Only one of these, namely $\frac{1}{2}\pi i$, lies inside our contour, and the residue there is

$$\frac{e^{ia\pi/2}}{\sinh(i\pi/2)} = \frac{e^{ia\pi/2}}{i\sin(\pi/2)} = -ie^{ia\pi/2}.$$

Hence

$$2\pi e^{ia\pi/2} = \int_{-u}^{v} \frac{e^{ax}}{\cosh x}\, dx + \int_{0}^{\pi} \frac{e^{a(v+iy)}}{\cosh(v+iy)}\, i\, dy$$

$$- \int_{-u}^{v} \frac{e^{a(x+\pi i)}}{\cosh(x+\pi i)}\, dx - \int_{0}^{\pi} \frac{e^{a(-u+iy)}}{\cosh(-u+iy)}\, i\, dy$$

$$= I_1 + I_2 - I_3 - I_4 \quad \text{(say)}.$$

Then

$$\left| \frac{e^{a(v+iy)}}{\cosh(v+iy)} \right| = \left| \frac{2e^{av}e^{iay}}{|e^{v+iy}+e^{-(v+iy)}|} \right| \leq \frac{2e^{av}}{e^v - e^{-v}},$$

and so I_2 tends to 0 as $v \to \infty$, since $a < 1$. Similarly,

$$\left| \frac{e^{a(-u+iy)}}{\cosh(-u+iy)} \right| = \left| \frac{2e^{-au}e^{iay}}{e^{-u+iy}+e^{u-iy}} \right| \leq \frac{2e^{-au}}{e^u - e^{-u}},$$

and so $I_4 \to 0$ as $u \to \infty$, since $a > -1$. Also, since $\cosh(x + \pi i) = -\cosh x$, $I_3 = -e^{i\pi a}I_1$. Hence, letting $u, v \to \infty$, we have $\int_{-\infty}^{\infty} [e^{ax}/\cosh x]\, dx = 2\pi e^{ia\pi/2}/(1 + e^{i\pi a}) = 2\pi/(e^{i\pi a/2} + e^{-i\pi a/2}) = \pi/\cos(\pi a/2)$.

9.23. The function $f(z)$ is holomorphic. On the arc α from R to $Re^{i\pi/4}$, $|f(z)| = R^{4n+3}|\exp(Re^{i\theta})| = R^{4n+3}e^{-R\cos\theta} \leq R^{4n+3}e^{-R/\sqrt{2}}$, and so $\left| \int_{\alpha} f(z)\, dz \right| \leq \frac{1}{4}\pi R^{4n+4}e^{-R/\sqrt{2}}$, which tends to 0 as $R \to \infty$. Hence, parametrising the line from 0 to $Re^{i\pi/4}$ by $z = (1+i)t \quad (0 \leq t \leq R/\sqrt{2})$, we deduce that

$$\int_{0}^{R} x^{4n+3}e^{-x}\, dx - \int_{0}^{R/\sqrt{2}} (1+i)^{4n+3}t^{4n+3}e^{-t(1+i)}(1+i)\, dt$$

tends to 0 as $R \to \infty$. Thus, since $(1+i)^{4n+4} = (-1)^{n+1}2^{2n+2}$, we deduce that $(4n+3)! = (-1)^{n+1}2^{2n+2}\int_{0}^{\infty} t^{4n+3}e^{-t}(\cos t + i\sin t)\, dt$, and equating real parts gives the desired result.

9.24. Let γ be the semicircular contour $\sigma(0, R)$, indented at the origin by a semicircle of radius r. The function $f(z) = (e^{iaz} - e^{ibz})/z^2$ is holomorphic inside and on the contour. On the outer semicircle Ω, $|e^{iaz}| = e^{-aR\sin\theta} \leq 1$, $|e^{ibz}| = e^{-bR\sin\theta} \leq 1$, and so $|f(z)| \leq 2/R^2$. Hence $\left| \int_{\Omega} f(z)\, dz \right| \leq 2\pi/R$, which tends to 0 as $R \to \infty$. Calculating the Laurent series of $f(z)$ at 0, we have $f(z) = (1/z^2)[(1 + iaz - a^2z^2 + O(z^3)) - (1 + ibz - b^2z^2 + O(z^3))] = i(a - b)z^{-1} - (a^2 - b^2) + O(z)$. Thus f has a simple pole at 0, with residue $i(a - b)$. Denoting the inner semicircle (in the positive direction) by ω, we know from Lemma 9.12 that $\lim_{r \to 0} \int_{\omega} f(z)\, dz = \pi(b - a)$. Thus, letting $R \to \infty$, we have $\int_{-\infty}^{-r} [(e^{iax} - e^{ibx})/x^2]\, dx - \int_{\omega} f(z)\, dz + \int_{r}^{\infty} [(e^{iax} - e^{ibx})/x^2]\, dx$. Letting

$r \to 0$ then gives $\int_{-\infty}^{\infty} [(e^{iax} - e^{ibx})/x^2]\, dx = \pi(b - a)$. The proof is completed by equating real parts.

Since $\sin^2 x = \frac{1}{2}(1 - \cos 2x) = \frac{1}{2}(\cos 0x - \cos 2x)$, it follows from the above that $\int_{-\infty}^{\infty} [(\sin x)/x]^2\, dx = \frac{1}{2}\pi(2 - 0) = \pi$. Hence $\int_{0}^{\infty} [(\sin x)/x]^2\, dx = \frac{1}{2}\pi$.)

9.25. Since the contour does not cross the cut $(-\infty, 0]$, we may assume that $\log z$ means the principal logarithm throughout. Since $a - e^{-iz} = 0$ if and only if $z = i \log a$, the integrand has just one pole, at $i \log a$. This does not lie inside the contour, since $0 < a < 1$ implies that $\log a$ is negative. Hence the integral round the contour has the value 0.

The contribution from the segment from π to $\pi + iR$ is

$$\int_{0}^{R} \frac{(\pi + iy)i\, dy}{a - e^{-i(\pi + iy)}} = \int_{0}^{R} \frac{(i\pi - y)\, dy}{a + e^y}\,,$$

and the section from $-\pi + iR$ to $-\pi$ contributes

$$-\int_{0}^{R} \frac{(-\pi + iy)i\, dy}{a - e^{-i(-\pi + iy)}} = \int_{0}^{R} \frac{(i\pi + y)\, dy}{a + e^y}\,.$$

Combining the two while letting $R \to \infty$ gives a contribution of

$$2\pi i \int_{0}^{\infty} \frac{dy}{a + e^y} = 2\pi i \int_{1}^{\infty} \frac{du}{u(a + u)} \quad (\text{where } u = e^y) \;=\; \frac{2\pi i}{a} \log(1 + a)\,.$$

$$(13.5)$$

The section from $\pi + iR$ to $-\pi + iR$ contributes

$$J = -\int_{-\pi}^{\pi} \frac{(x + iR)\, dx}{a - e^{-i(x + iR)}}\,,$$

and $|J| \le 2\pi(\pi + R)/(e^R - a)$. This tends to 0 as $R \to \infty$. Finally, the section from $-\pi$ to π contributes

$$\int_{-\pi}^{\pi} \frac{x(a - e^{ix})\, dx}{(a - e^{-ix})(a - e^{ix})} = \int_{-\pi}^{\pi} \frac{x(a - \cos x - i \sin x)\, dx}{1 - 2a \cos x + a^2}\,.$$

From (13.5), and by taking imaginary parts, we deduce that

$$-\int_{-\pi}^{\pi} \frac{x \sin x\, dx}{1 - 2a \cos x + a^2} + \frac{2\pi}{a} \log(1 + a) = 0\,,$$

and the required result follows since $x \sin x/(1 + 2a \cos x + a^2)$ is an even function.

9.26. The function $\pi \operatorname{cosec} \pi z/(z^2 + a^2)$ has poles at $\pm ai$ (and at every n). At each of the poles $\pm ai$ the residue is $(-\pi \operatorname{cosech} \pi a)/(2a)$. Hence $\sum_{n=-\infty}^{\infty}[(-1)^n/(n^2 + a^2)] = \pi/(a \sinh \pi a)$. That is,

$$\frac{1}{a^2} - 2\sum_{n=1}^{\infty} \frac{(-1)^{n+1}}{n^2 + a^2} = \frac{\pi}{a \sinh \pi a},$$

and so $\sum_{n=1}^{\infty}[(-1)^{n+1}/(n^2 + a^2)] = (1/2a^2) - \pi/(2a \sinh \pi a)$.

9.27. The function $f(z) = \pi \cot \pi z/(z+a)^2$ has a double pole at $-a$. The residue is $h'(-a)$, where $h(z) = \pi \cot \pi z$. Since $h'(z) = -\pi^2 \operatorname{cosec}^2 \pi z$, we have $\operatorname{res}(f, -a) = -\pi^2 \operatorname{cosec}^2 \pi a$. Hence, by Theorem 9.24, $\sum_{n=-\infty}^{\infty} 1/(n + a)^2 = \pi^2 \operatorname{cosec}^2 \pi a$.
With $a = \frac{1}{2}$, we have $\sum_{n=-\infty}^{\infty}[1/(2n + 1)^2] = \frac{1}{4}\sum_{n=-\infty}^{\infty}[1/(n + \frac{1}{2})^2] = \frac{1}{4}\pi^2$. With $a = \frac{1}{3}$, we have $\sum_{-\infty}^{\infty}[1/(3n + 1)^2] = \frac{1}{9}\sum_{-\infty}^{\infty}[1/(n + \frac{1}{3})^2] = \frac{1}{9}(4\pi^2/3) = \frac{4}{27}\pi^2$. With $a = \frac{1}{4}$, we have

$$\sum_{-\infty}^{\infty}[1/(4n + 1)^2] = \frac{1}{16}\sum_{-\infty}^{\infty}[1/(n + \frac{1}{4})^2] = \frac{1}{16}(2\pi^2) = \frac{1}{8}\pi^2.$$

9.28. The function $\pi \cot \pi a/(z^4 - a^4)$ has simple poles at a, $-a$, ia and $-ia$. At a and $-a$ the residue is $\pi \cot \pi a/(4a^3)$; at ia and $-ia$ the residue is $\pi \coth \pi a/4a^3$. Hence $\sum_{n=-\infty}^{\infty}[1/(n^4 - a^4)] = -(\pi/2a^3)(\cot \pi a + \coth \pi a)$. That is, $(-1/a^4) + 2\sum_{n=1}^{\infty}[1/(n^4 - a^4)] = -(\pi/2a^3)(\cot \pi a + \coth \pi a)$, and so $\sum_{n=1}^{\infty}[1/(n^4 - a^4)] = (1/2a^4) - (\pi/4a^3)(\cot \pi a + \coth \pi a)$.

Chapter 10

10.1. The equation $f(z) = 0$ has no roots on the positive x-axis, since $x \geq 0$ implies that $x^8 + 3x^3 + 7x + 5 > 5$. It has no roots on the y-axis either, since $f(iy) = (y^8 + 5) + i(7y - 3y^3)$, and $y^8 + 5 > 0$ for all real y. For $R > 0$ we consider the contour γ consisting of γ_1, the line segment from 0 to R, γ_2, the circular arc from R to iR, and γ_3, the line segment from iR to 0. Clearly $\Delta_{\gamma_1}(\arg f) = 0$. On γ_3 we have $\arg f = \tan^{-1}[(7R - 3R^2)/(R^8 + 5)]$. The quantity $(7R - 3R^2)/(R^8 + 5)$ stays finite throughout, is 0 when $R = 0$ and tends to 0 as $R \to \infty$. Hence $\Delta_{\gamma_3}(\arg f) \to 0$. As for γ_2, by choosing R sufficiently large we may use Rouché's Theorem to deduce that $\Delta_{\gamma_2}(\arg f) = \Delta_{\gamma_2}(\arg z^8) = 4\pi$. Hence there are two roots in the first quadrant.
We can establish that the roots are distinct by observing that there are no roots of $f(z) = 0$ on the line $\{re^{i\pi/4} : r > 0\}$. For $f(re^{i\pi/4}) = r^8 +$

$5+(1/\sqrt{2})(7r-3r^3)+(ir/\sqrt{2})(3r^2+7)$, and the real and imaginary parts can never be zero together. Then use two separate contours, 0 to R to $Re^{i\pi/4}$ to 0, and 0 to $Re^{i\pi/4}$ to iR to 0, and show that there is one zero inside each.

10.2. On the circle $\kappa(0,1)$, we have $|az^n| = a$ and $|e^z| = e^{\cos\theta} \le e$. Thus $|az^n| > |e^z|$ on $\kappa(0,1)$. The function $az^n - e^z$ has no roots on $\kappa(0,1)$ and no poles. By Rouché's Theorem, $\Delta_{\kappa(0,1)}\big(\arg(az^n - e^z)\big) = \Delta_{\kappa(0,1)}\big(\arg(az^n)\big) = 2n\pi$, and so there are n roots of $e^z = az^n$ inside $\kappa(0,1)$.

10.3. On the circle $\kappa(0,3/2)$, $|z^5| = 243/32$ and $|15z+1| \ge 15|z|-1 = 21.5$; thus $|15z+1| > |z|^5$. Hence there is no zero of the polynomial on the circle, and, by Rouché's Theorem, $\Delta_{\kappa(0,3/2)}\big(\arg(z^5+15z+1)\big) = \Delta_{\kappa(0,3/2)}\big(\arg(15z+1)\big) = 2\pi$. Thus there is one zero in $N(0,3/2)$.
On the circle $\kappa(0,2)$, $|z^5| = 32$ and $|15z+1| \le 15|z|+1 = 31$. Hence there is no zero of the polynomial on the circle, and, by Rouché's Theorem, $\Delta_{\kappa(0,2)}\big(\arg(z^5 + 15z + 1)\big) = \Delta_{\kappa(0,2)}\big(\arg(z^5)\big) = 10\pi$. Thus there are five zeros in $N(0,2)$. We deduce that in the annulus $\{z : 3/2 < |z| < 2\}$ there are four zeros.

10.4. Writing $f(z) = z^5 + 7z + 12$, observe that $f(-1) = 4 > 0$ and $f(-2) = -34 < 0$. Hence there is a real root between -2 and -1. Since $f'(z) = 5z^4 + 7 > 0$ for all z, there are no other real roots. There are no roots on the y-axis, since $f(iy) = i(y^5 + 7) + 12$ cannot be zero. If $|z| = 1$, then $|f(z)| \ge |7z + 12| - |z^5| \ge 12 - 7|z| - |z^5| = 4$, and so there are no roots on the circle $\kappa(0,1)$. Also, $|z^5 + 7z| \le 8 < 12$; so $\Delta_{\kappa(0,1)}(\arg f) = \Delta_{\kappa(0,1)}\big(\arg(12)\big) = 0$. Hence there are no roots in $N(0,1)$.
If $|z| = 2$, then $|f(z)| \ge |z^5| - |7z + 12| \ge |z^5| - 7|z| - 12 \ge 13$. Hence there are no roots on $\kappa(0,2)$. On $\kappa(0,2)$, $|7z + 12| \le 26 < |z^5|$. Hence, in each of the sectors $0 < \theta < 2\pi/5$, $2\pi/5 < \theta < 4\pi/5$, $4\pi/5 < \theta < 6\pi/5$, $6\pi/5 < \theta < 8\pi/5$ and $8\pi/5 < \theta < 2\pi$, $\Delta(\arg f) = \Delta\big(\arg(z^5)\big) = 2\pi$. The root in the third of those sectors is the real negative root already discovered, and the others are (respectively) in the first, second, third and fourth quadrants. All roots are in the annulus $\{z : 1 < | < |2\}$.

10.5. Let $R > 1$. On the circle $\kappa(0,R)$, $|z^n| = R^n = [1+(R-1)]^n > 1+n(R-1)+\frac{1}{2}n(n-1)(R-1)^2$ by the Binomial Theorem. Also $|nz-1| \le nR+1$. So we certainly have $|z^n| > |nz-1|$ if $1+n(R-1)+\frac{1}{2}n(n-1)(R-1)^2 \ge nR+1$, that is, if $-n+\frac{1}{2}n(n-1)(R-1)^2 \ge 0$, that is, if $R \ge 1 + [2/(n-1)]^{1/2}$. For any such R, $\Delta_{\kappa(0,R)}\big(\arg(z^n + nz - 1)\big) = \Delta_{\kappa(0,R)}\big(\arg(z^n)\big) = 2n\pi$, and so there are n zeros of $z^n + nz - 1$ in $N(0,R)$.

10.6. Suppose that f has no zeros in $N(0,R)$. Then $1/f$ is holomorphic in $\bar{N}(0,R)$. Then $|(1/f)(z)| < 1/M$ on the circle $\kappa(0,R)$, and $|(1/f)(0)| >$

M. Since f is not constant, this contradicts the Maximum Modulus Theorem. Hence there is at least one zero in $N(0, R)$.

Let $f(z) = a_0 + a_1 z + \cdots + a_n z^n$ be a polynomial of degree $n \geq 1$. Let
$M > |a_0|$. Since $|f(z)| \to \infty$ as $|z| \to \infty$, we can find a circle $\kappa(0, R)$
on which $|f(z)| > M$. Hence there is at least one root of $f(z) = 0$ in
$N(0, R)$. To show that there are n roots, use an inductive argument, as
in Theorem 7.11.

10.7. Let $f(z) = w = u + iv$. Since $|e^w| = e^u$, the existence of a maximum of
$\operatorname{Re} f$ on the boundary would imply the existence of a maximum of $|e^f|$,
and this is not possible, by the Maximum Modulus Theorem. If $\operatorname{Re} f$ had
a minimum value on the boundary, then $|e^{-f}|$ would have a maximum
value, and so this too is impossible.

Chapter 11

11.1. (i) For the first hyperbola, $2x - 2y(dy/dx) = 0$, and so $dy/dx = x/y$.
For the second hyperbola $2y + 2x(dy/dx) = 0$, and so $dy/dx = -y/x$.
The two tangent vectors are perpendicular.

(ii) The two parabolas meet where $u = k^2 - l^2$ and $v^2 = (2kl)^2$. For
the first parabola, $du/dv = -v/(2k^2)$, and for the second parabola
$du/dv = v/(2l^2)$. The product of the gradients of the two tangent
vectors is $-(v/2kl)^2 = -1$.

11.2. Observe that $g = f \circ h$, where $h(z) = \bar{z}$. The function h preserves
magnitudes of angles while reversing the sense, while f preserves both
the magnitude and the sense. Hence g preserves magnitudes of angles,
while reversing the sense.

11.3. a) Verify that $\partial^2 u/\partial x^2 = -\partial^2 u/\partial y^2 = 0$. From the Cauchy–Riemann
equations, $\partial v/\partial y = \partial u/\partial x = 1 + 2y$, and so $v = y + y^2 + f(x)$.
Hence $f'(x) = \partial v/\partial x = -\partial u/\partial y = -2x$, and so we may take $v = y + y^2 - x^2$. Observe that $f(z) = x + 2xy + i(y + y^2 - x^2) = (x + iy) - i((x^2 - y^2) + i(2xy)) = z - iz^2$.

b) $\partial^2 u/\partial x^2 = e^x \cos y$, $\partial^2 u/\partial y^2 = -e^x \cos y$, and so u is harmonic.
From the Cauchy–Riemann equations, $\partial v/\partial y = \partial u/\partial x = e^x \cos y$,
and so $v = e^x \sin y + f(x)$. Hence $e^x \sin y + f'(x) = \partial v/\partial x = -\partial u/\partial y = e^x \sin x$, and so we may take $v = e^x \sin y$. Observe that
$f(z) = e^x(\cos y + i \sin y) = e^x e^{iy} = e^{x+iy} = e^z$.

c) The calculations are more difficult here, but it is routine to verify
that $\partial^2 u/\partial x^2 = -\partial^2 u/\partial y^2 = (6x^2 y - 2y^3)/(x^2 + y^2)^3$. From the

Cauchy–Riemann equations, $\partial v/\partial y = \partial u/\partial x = 1 - 2xy/(x^2 + y^2)^2$, and so $v = y - x/(x^2 + y^2) + f(x)$. Hence $(y^2 - x^2)/(x^2 + y^2)^2 + f'(x) = \partial v/\partial x = -\partial u/\partial y = (y^2 - x^2)/(x^2 + y^2)^2$, and so we may take $v = y - x/(x^2 + y^2)$. Observe that $f(z) = x + iy - i(x - iy)/(x^2 + y^2) = z - i/z$.

11.4. Choose v so that $f = u + iv$ is holomorphic. Then $g : z \mapsto e^{-f(z)}$ is holomorphic, and $|g(z)| = e^{-u(x,y)}$. Then

$$e^{-m} = \sup\left\{|g(z)| : z \in \mathrm{I}(\gamma) \cup \gamma^*\right\}.$$

By the Maximum Modulus Theorem (Theorem 10.8), $|g(z)| < e^{-m}$ for all z in $\mathrm{I}(\gamma)$. Hence $u(x,y) > m$ for all (x,y) in $\mathrm{I}(\gamma)$.

11.5. a) $F(z) = \dfrac{z-1}{(1+i)z}$,   b) $F(z) = \dfrac{z-i}{z-1}$,   c) $F(z) = \dfrac{(-1+2i)z+1}{z+1}$.

11.6. By Remark 11.3, the local magnification is $|F'(\zeta)| = |(ad - bc)/(c\zeta + d)^2|$.

11.7.   a) The answer is not unique. Suppose first that F maps 1 (on the boundary of the disc D_1 to -1 (on the boundary of the disc D_2. Then map the inverse points -1, ∞ to inverse points ∞, -2 (in that order, so that the interior point 0 of the first disc maps to the exterior point ∞ of the second disc). This gives

$$F(z) = \frac{2z}{z+1}.$$

[Check the answer: $|w + 2| \geq 1$ if and only if $|z + 1| \leq 2$.]

b) Again the answer is not unique. We choose boundary points and map 1 to $3i$. Then we choose inverse points -1 and ∞ and map those to inverse points (reflections) $6i$ and 0 (in that order, so as to map the interior of the disc to the half-plane *above* the line). We obtain

$$F(z) = \frac{12i}{z+3}.$$

[Again we can check: $|z + 1| = 2$ if and only if $|(w - 6i)/w| = 1$, that is, if and only if w lies on the perpendicular bisector of the the line segment connecting 0 and $6i$.]

11.8. The transformation F must map the point 1 on D_1 to the point 0 on D_2. The points 0 and ∞ are inverse with respect to D_1, while the points $1/2$ and -1 are inverse with respect to D_2. Since $F(0) = 1/2$, we must have $F(\infty) = -1$. The Möbius transformation is completely determined by the images of 1, 0 and ∞:

$$F(z) = \frac{-z+1}{z+2}.$$

11.9.

$$(F \circ F)(z) = \frac{1 + i\left(\frac{1+iz}{i+z}\right)}{i + \left(\frac{1+iz}{i+z}\right)} = \frac{i + z + i - z}{-1 + iz + 1 + iz} = \frac{1}{z}.$$

Hence $F^3(z) = F^2\big(F(z)\big) = (i+z)/(1+iz)$, and $F^4(z) = F^2\big(F^2(z)\big) = z$.
Since z is real,

$$|w|^2 = w\bar{w} = \frac{(1+iz)(1-iz))}{(i+z)(-i+z)} = \frac{1+z^2}{1+z^2} = 1,$$

and so $w = F(z)$ lies on the circle $\{w : |w| = 1\}$. From $F(-1) = -1$,
$F(1) = 1$ and $F(0) = -i$, we deduce that the image of the line segment
from -1 to 1 is the semicircle $\{e^{i\theta} : -\pi \le \theta \le 0\}$. The image under F^2
is the union $(-\infty, -1] \cup [1, \infty)$ of intervals on the real line, and the image
under F^3 is the upper semicircle $\{e^{i\theta} : 0 \le \theta \le \pi\}$.

11.10. Observe first that $\operatorname{Re} z \ge 0$ if and only if $|(z-1)/(z+1)| \le 1$. So
the image under $z \mapsto (z-1)/(z+1)$ of the right half-plane is the disc
$\{w : |w| \le 1\}$. By contrast, the real axis maps to itself, and, for all real
p, q,

$$\frac{p+qi-1}{p+qi+1} = \frac{\big((p-1)+qi\big)\big((p+1)-qi\big)}{(p+1)^2+q^2} = \frac{(p^2-1+q^2)+2qi}{(p+1)^2+q^2}$$

has positive imaginary part if and only if $q > 0$. Thus the positive half-
plane maps to itself, and so the first quadrant maps to the semicircle
$\{w : |w| \le 1, \operatorname{Im} w > 0\}$. The square function maps this semicircle to
the complete closed disc $\bar{N}(0,1)$. So the image of Q under $z \mapsto [(z-1)(z+1)]^2$ is $\bar{N}(0,1)$.
As for $z \mapsto (z^2-1)/(z^2+1)$, the square function maps Q to the upper
half-plane $\{w : \operatorname{Im} w \ge 0\}$, which then maps to itself by $w \mapsto (w-1)/(w+1)$.

11.11. First, $z \mapsto z^4$ maps E to $E' = \{z : |z| \ge 1\}$. A suitable mapping is
$z \mapsto 1/z^4$.

11.12. First map Q to the first quadrant $Q' = \{z : \operatorname{Re} z > 0, \operatorname{Im} z > 0\}$ using
$z \mapsto (\pi/2) - z$. Then multiply by $e^{-i\pi/4}$ to transform Q' into $Q'' = \{z : -(\pi/4) < \arg z < (\pi/4)\}$. Then square to obtain the half-plane
$\{z : \operatorname{Re} z > 0\}$. So a suitable mapping is

$$z \mapsto \left(\frac{\pi}{2} - z\right)^2 e^{-i\pi/2} = -i\left(\frac{\pi}{2} - z\right)^2.$$

Bibliography

[1] Lars V. Ahlfors, *Complex Analysis*, McGraw-Hill, 1953.

[2] Ian Anderson, *A First Course in Combinatorial Mathematics*, Oxford University Press, 1974.

[3] A. F. Beardon, *Complex Analysis*, Wiley, 1979.

[4] A. F. Beardon, *A Primer on Riemann Surfaces*, London Math. Soc. Lecture Note Series, no. 78, Cambridge University Press, 1984.

[5] R. V. Churchill, *Complex Variables and Applications*, McGraw-Hill, 1960.

[6] William Feller, *An Introduction to Probability Theory and its Applications* Volume 1, Wiley, 1957.

[7] Kenneth Falconer, *Fractal Geometry: Mathematical Foundations and Applications*, 2nd Edition, Wiley, 2003.

[8] G. H. Hardy, Sur les zéros de la fonction $\zeta(s)$ de Riemann, *Comptes Rendus* **158** (1914) 1012–1014.

[9] John M. Howie, *Real Analysis*, Springer, 2000.

[10] G. J. O. Jameson, *A First Course in Complex Analysis*, Chapman and Hall, 1970.

[11] J. J. O'Connor and E. F. Robertson, History of Mathematics web site `www-history.mcs.st-and.ac.uk/history/`

[12] Heinz-Otto Peitgen, Hartmut Jürgens and Dietmar Saupe, *Chaos and Fractals*, New Frontiers of Science, Springer-Verlag, New York, 1992.

[13] Murray R. Spiegel, *Advanced Calculus*, Schaum's Outline Series, McGraw-Hill, 1963.

[14] E. C. Titchmarsh, *The Theory of Functions*, Second Edition, Oxford University Press, 1939.

[15] E. C. Titchmarsh, *The Zeta-function of Riemann*, Cambridge Tracts in Mathematics and Mathematical Physics, No. 26, Cambridge University Press, 1930.

[16] E. T. Whittaker and G. N. Watson, *A Course of Modern Analysis*, Fourth Edition, Cambridge University Press, 1927.

Index